Praise for Ward Mayfield

Lessons From Uncle Sam is a unique testimonial to the resilience of today's OIF/OEF veteran—candid and irreverent as only one who has been there—a must-read!

— William W. Hansen, *President*, Cavalry Creek Consulting Group

Ward's vivid recounting of his missions place you in the vehicle next to him for the best accounting of route clearance operations to date.

— Michael Trofinoff, Colonel, Engineer, US Army

Lessons From Uncle Sam

Ward Mayfield

TACTICAL 16
PUBLISHING

Lessons from Uncle Sam

First Edition

Published by Tactical 16 Publishing

Colorado Springs, Colorado

www.Tactical16.com

ISBN: 978-1-943226-70-2 (paperback)

Contents

Chapter 1

Who the Hell Am I?

Who the hell am I? Nobody, really. I'm just a storyteller. My whole life I've enjoyed regaling people with tales of my own brand of dumb luck for a laugh. As a kid, I made up fantasies about all sorts of things to tell folks: dragons, mobs, aliens, dinosaurs, the usual. As I got older I realized that, if told well, life presented all sorts of great moments that didn't need monsters or time-traveling space rangers to entertain the listeners. All that was necessary was honesty and the inclusion of embarrassing, human moments. Throughout elementary, middle, and high school, getting attention by way of clowning around and over-the-top antics was my normal. None of it was malicious in nature. I just endlessly questioned the universe, curious about everything in it. Which often did, and still does, get me into heaps of trouble.

Beginning at the age of maybe four or five and finishing with being kicked out of high school at eighteen, I:

- Was permanently banned from a kiddie amusement park in Texas for running amok, riding rides without tickets or adults, and stealing cotton candy.

- Broke the family TV by sticking an industrial-sized magnet on the screen, permanently changing the colors.
- Tore holes in the bottom of antique furniture because my dad had passingly said, "If you can figure out how to get to the change, you can have it."
- Flooded our entire home because my toy dinosaurs drowned in a great storm in the upstairs bathroom sink before school and the water was left on.
- Plugged a fork into an electrical outlet just to know what it felt like.
- Nearly killed myself by taking apart a plugged-in clock radio while showering.
- Got expelled from middle school for stealing a car to go somewhere else for lunch.
- Crashed a friend's car into three other cars five days before my sixteenth birthday.
- Got suspended for forty-five days for the distribution of narcotics my junior year.
- Got expelled from high school just weeks before graduation for public intoxication and vandalism at the senior dinner dance.

That isn't even a fraction of the childhood mayhem I caused, and a lot of that is self-explanatory. A few things require further explanation. My buddy lived across the street from Midvale Middle School with his older brother, whose car we borrowed one day to go for tacos. His brother worked nearby and happened to walk home for lunch while we were out. He noticed his car was missing so he called the police. I wasn't the driver for our little joyride and no charges were filed, but since we had left school grounds and caused a debacle that involved police, I was immediately expelled from school.

The narcotics I was caught distributing were dilating eye drops stolen from an ophthalmologist's office. A bunch of us were curious but too afraid to put drops in our own eyes. I volunteered to put drops in all of our eyes, in full view of school security cameras. As for the car accident, a friend had driven me to my house for lunch and heavily sampled my parents' whiskey. When we discovered he was too drunk to drive back, I volunteered to do it. Having zero experience behind any wheel I took the final turn back to school way too fast and clobbered three of the four vehicles in the turn lane on Hillcrest High Drive, totaling all of them. The crash was pretty epic and amazingly, no one was seriously hurt. My entire high school watching while the cops detained me and my friend on the street corner was less epic and hugely embarrassing. Calling my vacationing parents in Vegas was worse.

The clock radio, magnet, fork in the socket, and antique furniture destruction are prime examples of the intense compulsion I feel to understand things at their most basic levels. When I'm stumped, I'll poke, analyze, and disassemble something until I get it. Then I feel really bad that whatever it was is now most likely a broken, but still unique, paperweight.

I had already joined the army the summer before my senior year on July 9, 2001. Hadn't gone to basic yet; just signed the dotted line. A huge scholastic slacker, I had completed only enough make-up packets to barely earn my diploma. On the train headed to the last big dance for the graduating class near the end of the school year, I celebrated my hard-won success by drinking a sixteen-ounce water bottle filled with whiskey. Shortly after my arrival at the dance, I was discovered to be very intoxicated and carted right out of the ballroom. Outside, I was informed by police and school officials that I wouldn't be walking down graduation lane during the ceremony or even allowed to attend. Drunken and indignant, I responded poorly. Thanks to a tantrum culminating in striking someone with a park bench outside the downtown Salt Lake Hilton, I was expelled. The last time I was able to walk the halls of Hillcrest High School was with a police escort turning in my books, paying any outstanding fines, and receiving a completely unsigned yearbook. Since I had yet to attend training and now had legal issues, the army released me from service a few days later.

Realizing I was veering toward a headlong crash into destitution, I resolved to get my shit together using a commitment I had already made. A little over six months later, my plea in abeyance and community service were finished. On December 19, 2002, I re-entered the army with the actual drive to succeed at something. Along with my attitude, my stories were changing. They were still just my way of relating to others, but people seemed to listen a lot more intently than before. Not like standing on a milk crate at the mall proselytizing, but around a campfire or bonfire story-swapping. Frequently, I'd get told to write a book about my experiences. I didn't give it any real consideration until I was thirty years old going through retiree out-processing in Fort Carson, Colorado. I was sitting in a plastic chair somewhere in the middle of a fifty-person-long line, half optimistically and half despairingly pondering: *Someday, I may not remember any of this.*

Among the small myriad of injuries given as reward for a few hard knocks to the noggin thanks to bombs, I have Traumatic Brain Injury, or TBI. TBIs are diagnosed as mild to severe with a few other levels in between. Depending on the day, I'm somewhere between the mild and moderate areas of functionality. This means most days I look and talk just fine—a tic or two here, a stutter there, almost imperceptible. On terrible days, grasping the process of starting a car can be more complicated than is possible for me. Though those days aren't frequent, they are frightening. Having to deal with days where understanding is more difficult or painful than it should be isn't fun, but at least I know why it's happening. One of my biggest fears is that one random morning, I'll wake up in pain and not know why anymore. Same goes for everything else. With three tours—two to Iraq and one to Afghanistan—more missions completed than there are days in a year, and at least a baker's dozen IED strikes survived, my mind is no longer in "like new" condition. The possibility of eventually losing it altogether is very real. Thus, putting my memories on paper began sounding like a more and more reasonable idea. Nowhere near all of them are here, of course. What is here isn't always pleasant. But these pages are filled with pieces of time that had tremendous and lasting impact on my life and how I view the universe.

This isn't a book filled with the right way to get things done though, not even close. I wasn't Special Forces or a superhero wearing underwear outside my pants saving babies with my eye lasers. And while many lessons were taught more than once, these are the memories and moments that play in my head over and over the most.

These pages are the army split seconds that have etched themselves into what shapes who I am and what I choose to value going forward with my life. Since this book was written to remind future me of just who the hell I was, I say it works.

So this is a story to myself, written by a short and "slightly special" retired military troll man child. If at some point you (anyone not me) feel like I'm yelling or getting preachy, it's just me yelling at me. Family and friends having access to my noodle is an appealing concept as well since I can't really talk about most of this stuff without a lot of blubbering, a bunch of water coming out of my face, and a slightly heightened likelihood of seizing. The long and short of it: I just want the people in my life to understand who and why I am. I'd also like one of those people to be future me. Hopefully this book accomplishes that.

Writing started in earnest in fall 2016 and here I am in fall 2021 looking to put the final words down. Writing this cluster of pages went in furious spurts followed by long periods of doing anything other than writing. Over the course of five years this book has blessed me with a few bouts of depression, some seizures, and a handful of pass-out falls, one culminating in a hospital visit. Simply put, this wasn't as easy to do as initially expected. If this never gets beyond an unpublished, three-ring binder on my bookshelf, that's fine. If there are parts that are confusing or seem as if chunks of time are missing, that's fine too. It more closely resembles my mind. I'm not a chronological archivist master and this isn't an ordered timeline of events with every detail annotated. I tried that and it nearly broke my brain. I still wanted and needed to finish this book honestly and in my voice. So that's what I did, I think. I may have actually had eye lasers at one point. Either way, here I am.

I thought long and hard about memorializing those who died serving alongside me with some kind of list. When I got to the other side of that inner argument, my conclusion was that this book would be many thousands of pages longer if I gave honest tribute to each and every loved one. With a volume that large, I fear my personal lessons would be lost among such greatness. Therefore, to keep this book short and simple enough for future me, and to avoid the feeling of cheapening their deaths, I dedicate the entirety of this overlong rant to all of my fallen family, friends, and brothers and sisters in arms who will never have the opportunity to read it.

An enormous and eternal thank you to everyone who encouraged me to pick the pen back up every time I dropped it. Especially you, Momma.

P.S.: If you happen to not be me, or even future me, from the very side of my insane little heart, thanks for peeking inside my brain.

Chapter 2

Semper Tentare

Semper Tentare means "Always Try" in Latin. It is the motto of the 321st Engineer Battalion of the United States Army Reserve, my battalion. When you first join the Reserves, you can still attend monthly drills before going to basic training. You are called a Pre-Basic or Pre-B and essentially you can legally do no soldiering at all, but attendance is highly encouraged as it can help you get familiar with a few things before going to your Initial Entry Training site. I was assigned to Charlie Company (C CO) out of Ogden, Utah. I was seventeen years old, so when I arrived at the unit and learned of my new motto, I didn't see it as cool enough for a combat soldier. It wasn't the word "always" that irked me. It was "try." I mean, come on. There had to be something more battle-appropriate than "try." Something more awesome like, "Blow 'em to pieces!", "Eat evil people!", or "Fuck bad guys!" I don't know, something more and bigger and louder.

The first few years in, I continued to think that way. Then came deployments. Over the course of several long and difficult years spread over several countries, I learned that I was 100% incorrect. There is nothing more ferocious, steadfast, or inspirationally powerful than a person who never quits. Through anything and everything thrown at them, they just keep going. We've all seen or heard a campy version of that very statement, so it isn't exactly a new concept. Think of all the silly inspirational posters you see plastered on school walls and in office building break rooms. You know them, with the pictures of really cool stuff bordered by phrases like, "Never give up!" "Winners don't quit!" Yada yada. But when I continuously saw that determination and resilience in war it was something else entirely. Always trying means occasionally failing. And in that environment, you don't lose the championship, you don't get passed over for promotion, and you very rarely get a second chance. In combat, failure can, and often does, go hand in hand with grievous, life-changing injury or death. But a lot of the time the person who fails isn't the one who pays the cost. It's someone we were supposed to be protecting, our brothers in arms or the civilian populace. In more than a decade with the army I was party to great loss and failure, but also to powerfully fortifying successes. For me the failures were the most important lessons because they taught me to learn from those failings and helped me achieve the successes that came later.

Over time and through some of the most difficult days and nights of my life so far, *Semper Tentare* came to mean more and more. It is a simple, two-word phrase with a whopping five syllables (only three in English). Yet over and over again, I have seen people struggle to the brink of collapse or death and somehow still finish the job. Whether you are tall, short, skinny, fat, dark-skinned, light-skinned, educated, uneducated, penis or vagina between your legs—it simply does not matter. I have seen folks of all walks of life do amazing shit. One night in particular taught me about the true nature of always trying, and the heavy cost of failure.

It was 2006 and I was on my second deployment doing route clearance in Ramadi, Iraq with Bravo Company (B CO) 321st Engineers out of Hayden Lake, Idaho. A sister unit to mine, C CO. Our job was simple in description: clear the route—roads both paved and dirt—so other units could travel safely and perform their operations. Clearing a route meant that our team would drive in front of the landowner's team and either remove, or get hit by, any explosives in the road. That was the job. Before a unit would conduct its mission, we went in ahead and cleared them a path. Which is why my battalion came to be known as Pathfinder. Route clearance is commonly regarded as one of the most dangerous jobs of the Iraq and Afghanistan wars. Having done it in both countries, I'm inclined to believe that. Most dangerous, I don't know, but certainly among the most.

The mission started at 1700 so line up was at 1530. This is where we get our trucks ready. We wash the windows, mount our weapons, get radios up and running, put the mission's route into the GPS, make sure we have food and water, brief the night's route, don our armor, and pray. Monty was truck commander in the back of truck. Clark was assistant driver in the front passenger seat. Walt was gunning the 240B (7.62 millimeter machine gun, back of truck, when not up in turret). I was the driver. That night, as all nights up to that point in the deployment, we were working for a landowner unit called 136th Infantry, Company Bravo. We just called them "136" and that night we were working for Third Platoon. Once our trucks were up and everyone was briefed on the night's operation, we rolled on over to 136's motor pool to link up and do one final unified briefing. The motor pool is where a unit can park and work on all their equipment. We met at theirs because they were Abrams and Bradley operators. Those are tanks and tracked fighting vehicles, the big boys. They required much more space than our trucks. So we'd meet at their house, do our final, and roll out Trooper Gate.

We worked for 136 all the time. I can't remember how many we did with them exactly, but the guys in my platoon seemed to agree it was around seventy-seven. Not five-day blocks called work weeks with weekends in between. Seventy-seven days, seventy-seven missions. Almost all of them were really long with primary, secondary, and tertiary routes to clear. In the beginning the only days we were given off were when we had trucks disabled by IED hits. Eventually the army designed mandatory down days, called "maintenance days," so we got to know the 136 guys fairly well. It wasn't on a "Hey Bob, how's the wife and kids?" basis though. Hell, the only guy whose name I can even remember from that entire company was Shaffer. And I can't remember his first name. We weren't close, but I knew him. Physically worked together on a near-daily basis. Shaffer and his crew were one of three Bradley Fighting Vehicles with us that night.

Bradleys are tough in their own right, with some awesome armaments. A 7.62 laser-guided machine gun, a 25-millimeter, high-explosive main gun, and a few other explosive goodies. They're also equipped with a material called reactive armor. So if a rocket-propelled grenade (RPG), mortar, or regular old rocket hit them, the reactive armor would explode and push the incoming blast away from the vehicle. It kind of mitigates the damage of an explosion with another explosion. Super cool in theory. Somehow the armor always fell the fuck off the vehicle. Always.

"Hey, the next corner is really tight."

"How do you know?"

"Reactive armor box there on the ground."

"Shit."

We'd have to pick up reactive armor boxes on the side of the road so often that, eventually, we started leaving them on the ground. We had three Bradleys in tow that night; they would follow directly behind our last vehicle or Vic, my truck.

With everybody briefed, prepared, and loaded up in their vehicles, we rolled out the gate. Our mission was to take Railroad and clear a few inner city streets in preparation for a survey operation of that neighborhood the following morning. The army isn't very innovative with street names. Railroad was an actual elevated railroad bridge that was one of our entry points into downtown Ramadi. On either side of the tracks was a drop ranging fifteen to twenty-five feet into sand or water. These trucks sink fast and can only be picked up by specialized heavy-duty wreckers. If we fall, we're in deep doodoo. Keep in mind it was November and got dark early, and route clearance was, more often than not, the slowest damn job in the world. Depending on the road type, we drove between one and five kilometers an hour. It took us a while, and by the time we got to Railroad the sun was already down.

We'd only been in Ramadi for a little over a month and we had not completed many missions. This was about our thirty-fifth or fortieth mission of the tour. Even though we'd driven on the road multiple times, it was still crazy scary. It was a pitch-black night. We didn't roll in the daytime. There were a million reasons for that but the reason that made most sense to me was Ramadi was a big and populated city.

We didn't want to be finding, or worse yet, getting found by IEDs while there were hundreds of people walking about conducting their daily business. So nighttime missions it was. Because of a tactical standard operating procedure, only the first couple of trucks ran with their headlights on. We had serious light setups: high-powered headlights, light bars, cop lights (spotlights with a movable beam), and a slew of aircraft landing lights (about ten to twelve per truck). In full light mode it was as if someone had turned on the sun. The back of the clearance team ran in blackout.

To drive in blackout, we wear night vision goggles, which illuminate everything, instantly. It feels just a little less bright than full daylight when you're looking through them. That's how clear the picture can be if you do your focusing correctly. But depth perception goes straight to hell. If you are picturing the binocular style, we didn't have those. We had the updated, upgraded version, a monocle. The military labeled them with some fancy acronym. I just called them the monocle because it made me think of the little Monopoly guy. We used the monocle to use both eyes to look at a target through different light sources. One eye uses natural light, just your naked eye. The other eye has your depth perception, fuck-up monocle. But you can see super clear. You had depth perception in one eye but couldn't see shit. The other eye had a clear image, but you couldn't tell how far away anything was. That's how it worked.

Vehicle spacing rules called for twenty-five to fifty meters between vehicles. This meant trucks running in white light were often almost two football fields away from mine. Imagine you're driving downtown in your car and your lights are off. Then imagine all the city lights are also off. Ramadi completely powered down at night. Unless you had a generator of your own, there were no lights. If there were, they were just tiny, little, far-off blips. You know, a front porch is illuminated a few blocks down. One intersection has a single working bulb. Someone has a candle burning in a window on the third floor of a building. It's fucking dark... and you're looking for bombs.

In addition, your car is wide, tall, and weighs over ten tons. Imagine monster trucks mounted with guns and armor. I'm five feet six and the tires of our biggest rig, the Buffalo, came up to my neck.

They're huge vehicles. We used to joke that my truck, the RG-31, looked like a Range Rover, but instead of Range Rover "XL," this model would be called the Range Rover "Holy Shit." I had never seen vehicles like that before. The usual things I thought about in regard to military vehicles were Humvees and tanks. That's it. Then I got into an RG and realized that it had everything the civilian world had. Only army vehicles are almost all black, olive drab green, sand brown, or some camo pattern containing those colors. Most of them were clunky pieces of shit. So the RG, Buffalo, Husky, and Cougar were super-impressive army trucks to me.

The rear RG was used primarily as a gun truck and we didn't get into too many firefights, so most of the time, we scanned around with night vision goggles. You scan hard looking for something, anything out of place from yesterday or the last time this road was on a mission route. *Is that dirt a little funnier color than the rest of the dirt around it? Was it disturbed recently? Is that shiny thing on that rock just some kind of reflective mineral in the rock itself or am I catching the glint of a wire? Is that a crack in the concrete or is it a specifically cut "lid" to a cutout-style IED with dust over the top of it?* Every single thing was examined. And I could go on for days—the bottom line is that we looked for it all. It was mildly stressful and not awesome like the army commercials with the catchy slogan suggested.

At the end of Railroad we could either turn right into Combat Outpost Iron or left onto Spartan Street. We took a left. Spartan was a very wide street, maybe half a football field, and it wasn't really a road, so to speak. Houses lined both sides of the expanse from nut to butt. Most houses were surrounded by some type of concrete wall high enough that we couldn't see over it. There were small patches of sporadic concrete or asphalt slightly larger than sidewalk width right up against some of the house walls. But the majority of the Spartan was dirt, mud, loads of garbage, grass reeds, or Shit Creek. Yes, Shit Creek. That was the actual on-a-military-map-printed-in-top-secret-operation-centers name of a putrid, green body of water. If you threw something like a bottle into the creek, it just sat on top of the thick layer of greenish-brown foam covering a poop ditch that amazingly did have liquid running through it. This thing was disgusting. And actually, my friend Andrew had to have surgery on his heart because of an infection he got while we were in said creek digging out some mines on another mission a few months later. He's the only one of us that got really ill, but everybody that got in the water had varying degrees of *Oh-dear-God-let-it-end-diarrhea* and/or exorcism vomiting. Our job was not always glorious.

We rolled down the various inner city streets of our route for about three hours without incident, then all commo on the lead truck went out. What that meant was their radio took a giant, fat shit. Unacceptable. The lead truck must always have communication—it is the one in front of the team identifying threats. It is the one telling the Buffalo, which has the interrogation portion of our platoon, in detail what the threat appears to be and then shines a laser on the area of interest. The lead truck had to have commo. Which truck didn't? The rear truck, mine. After we reached somewhere safe-ish, our only option was combat cross load. One crew egresses, or exits in haste, from the downed vehicle and switches out with the crew in the functioning vehicle. The engine of the blind vehicle operated just fine, so the choice wasn't a difficult one. With no commo or GPS, the lead RG could not be Mission Capable. But now my truck no longer had a radio, a BFT (Blue Force Tracker, military GPS), or an ECM, which stops IEDs from being detonated remotely, for example, with a cell phone. And we drove without lights. To spell it out simply for the non-military readers: we would now be completely blind and mute for the rest of the night. Even if the powerful, white light was needed, the best we were going to be able to muster was two feeble headlights.

We continued about thirty more minutes before Murphy's law, "Whatever can go wrong, will," took tighter hold of our evening. The new lead truck, the one I was previously driving, completely lost all transmission capability, meaning it could not shift gears so moving forward was very iffy. Their commo was still up but they could not move as needed. This was unknown to us in the rear truck because our commo was still down. After we blindly waited for the first few agonizing minutes to pass, we decided that Walt could be our commo. During stops he would be in the turret yelling to the gunner of the vehicle in front of us to find out what was going on. This was to be our only form of communication with the outside for the rest of the night. The lead truck and wrecker crew took maybe fifteen to twenty minutes to get the truck to at least roll forward, but reverse was going to be impossible for the rest of the mission.

During this debacle, I can't recall if it was somebody in our platoon, one of the Bradleys, or even one of the Abrams tanks who provided twenty-four-hour overwatch of particularly nasty places in town, but a report came down that we were being shadowed by another car on an adjacent street. According to the rules of engagement you were not permitted to use lethal force unless it had already been used against you or you had a crystal-clear indication that it was about to be used. That means if someone tells you that they're going to kill you, you can't do anything. On the other hand, if someone standing in front of you points a gun at your face and you absolutely believe they're going to shoot you, then you can engage them. So since the car was only keeping pace with us, we just kept tabs and went about our business of waiting on the front truck.

At first they weren't doing anything out of sorts except shadowing us, but Ramadi was not the place to fuck around. Throughout the city there were dozens of rubble piles where houses and buildings once stood. Almost every home had been redecorated with bullet and blast holes—Ramadi was an actual war zone. The car pulled ahead and seemed to no longer be interested in us. They drove to an intersection a few hundred yards from the lead truck and got out of the car, about five or seven people total. When they started unloading stuff from the trunk, it took us a minute to figure out what they were doing because they were so far away. The minute we saw them dinking around in the road, however, we knew they were planting a bomb. But we had to have absolutely positive identification.

A few minutes later we finally we got a good angle and could see they were putting artillery rounds in the ground to make an IED. Once we had that visual, it came down from higher that we were going to sympathetically detonate the IED (blow it up by hitting it with an explosive of our own) and get rid of the bad guys at the same time. As I said, Abrams tanks were parked all over Ramadi as overwatch for particularly ugly parts of town. There happened to be one in the vicinity. So a tank crew rolled on over, confirmed identification on the group and sent a 120-millimeter main gun round down to say hello. An Abrams main round is bigger than your forearm, much bigger. It's a big-boy bullet, and it explodes. One round was all it took to blow them all apart.

"Yay. Victory for the good guys. Back to work."

As soon as the lead truck crew was able to finally roll forward again, some idiot with grenades jumped into the middle of our formation and got himself obliterated. Just popped out from behind a wall with grenades in his hand and on a bandoleer around his chest clearly not trying to turn them in to us.

"Great plan, pal. Run out to throw grenades. Why did you run out?"

I still don't know why he did it but jumping out from cover got him shot to pieces before he could even toss one grenade. We had to drop a couple of our boys on the ground to police up the grenades from the dead guy. We left the body because usually there's a loved one nearby. We didn't mistreat the bodies, but we did remove intel, weaponry, and dangerous objects from them.

While this was going on, the lead truck made it all the way up to the devastation of the tank round, and it became very apparent that these people had, in fact, been putting a sizable IED in the ground. We got most of the IED ordnance to blow. It definitely killed all of the folks who had been placing it. Some were children. Not like five year olds or babies. But the group age ranged from about ten to thirty-five. First and foremost we had to make sure that the ordnance was gone and we rendered it all safe. We could not allow anyone else to come out and get blown up on our side or theirs. So the bodies had to be moved. Kudos to our arm operator that night. The arm operator is the guy who runs the hydraulic interrogation arm on the Buffalo that we use to dig IEDs out of the ground. It has a thirty-foot boom, five eighteen-inch tines and one two-foot-long, very thick tine on the back. We called the five tines fingers, and the one big tine the thumb.

It would have been simplest to spear the bodies with the thumb and throw them into a pile because they're dead combatants, right? But they weren't just combatants to us. They were mangled and dismembered men and children strewn about the street with viscera, blood, and horror everywhere. Our arm operator, God bless him, took a very long time setting up angles and approaches to get the arm to slide underneath the bodies without cutting them up. One by one, piece by piece, he picked them up and put them down gently on the side of the road or on the sidewalk if he could get them there. This did not take three- to-five minutes. It could have taken twenty, but it took an-hour-and-a-half because we wanted it done respectfully. That's something I won't forget.

It took ages because as we advanced, we were inching forward still trying to investigate the presence of secondary devices and unexploded ordnance. In all that time, my truck had moved, maybe one hundred meters. We were in the heart of the dangerous part of Ramadi, one of them at least. If you went to any part of Ramadi and asked a local, "Which are the worst streets, you know, the streets American soldiers should never walk down?" the answer would always be, "Everywhere."

We had a color scheme to indicate road safety. Green was good, red was dangerous (high probability of attack), and black routes. Black routes were impassable. No one drove down black routes. They would get destroyed. Or at the very least their truck would end up deadlined, meaning their vehicle would be inoperable either from enemy attack or obstacles they couldn't get past. Nothing but black roads surrounded us. But we were determined that while we were there, we were going to make those roads green. That was part of our mission that night. It was the big push to make the roads safe again.

Route clearance is like chess: easy to play, super hard to master. The problem is, if you don't play really well every time, someone dies. The grenade guy was handled, and everyone was waiting for the arm operator to finish up front. The Bradley behind us closed the distance but got a little too close. We were lazy in our truck about making sure we kept at least some semblance of spacing. Their barrel wasn't up our butt or anything, but they were far too close—bumpers within three meters or about ten feet. You'd think that would be far enough. In the world of bombs and explosions, however, I was about to learn that ten feet was indeed far too close.

By now it was late into the night. Close to midnight; I'm not sure exactly. But we'd been on this two-hundred-meter stretch for a while and the radio was now silent. My truck couldn't hear anything anyway, but no one was chatty. We were kept in the loop because Walt would get in the turret and get updates yelled over from another truck's gunner. Everyone was just waiting patiently for the arm operator to finish his awful-ass task.

The Bradley was probably repositioning again. I know we weren't moving. But out of nowhere, silence was shattered. I had never heard anything so deep in my life. It sounded like it came from the earth's molten core. That's how low the boom sounded. At that moment, I understood when bombs go off below ground, they sound a lot different than when they blowup on the surface with nothing on top of them. I'd learned about it at combat engineer school years earlier and from many times blowing things up at demo ranges. But that was my first up-close and personal lesson. Even today whenever I hear it, my heart drops instantly. I can describe this sound in different ways. It's a very muffled, but very loud, deep, bellowing sound. It's a tough thing to describe. It's a loud-as-shit, scary bomb noise followed by crashing as you, your truck, and road debris smash back into the planet. Insert the word muffled in there somewhere. Loud or quiet, whenever the sound is replicated near enough to grace my ears, it sends an icicle of fear down my spine.

The blast picked up my ten-ton truck about a foot and a half and moved us a few feet forward. As I said earlier, I was the driver. We were in park. I had my foot on the damned brake, and we still moved. Just shoved us forward. We looked around for a second while the dust cleared.

"What the shit just happened? Huh? Are we hit? Are we dead?" We weren't sure if we got hit with some little fart of a bomb or what. It didn't occur to us until the smoke and fire showed up that we didn't get hit at all. We were just so close to the Bradley that we thought we were.

Maybe we acted tactically, but it was all automatic and completely frantic at that point. We did our "zeros, fives, and twenty-fives." We visually checked the immediate area around our vehicle to ensure it was clear of secondary devices before exiting or even moving the vehicle. Zero, five, and twenty-five represent distances in meters. Clear around the door, zero. Clear just past the door, five meters. Clear to the nearest cover, twenty-five meters. Then get out of, or move the truck, to get to wherever it is you need to go. Monty went back to maniacally attempting to get the radio working again, because a fucking bomb just went off and we still needed to alert the platoon about who had been hit. Clark temporarily kind of lost his mind and just stared at the floor yelling about dropping and losing his flashlight. The back half of the truck was covered in dirt and debris, so we couldn't see shit through the rear windows. Walt popped up into his turret to get a situation read. Unsure of what I should do, I just put the truck in reverse and repositioned to where we were before the blast so we could provide cover. We still couldn't get the radio to do anything and didn't really know what was going on outside. We hadn't gotten hit by the IED. But looking back, if it was enough force to move our truck, we were probably a little more than just rattled right then. Not having a radio only made it worse.

Walt popped down from the turret every few seconds to give us status updates. His head furiously bounced around up and down to check on the Bradley, yell to other trucks to relay information, and update the three of us still below in the dark and dust.

"I'm fucking going. We... we gotta go because—I don't know what's going on, but fuck the..."

He stopped for a nanosecond to breathe.

"The vehicle's on fire—the fucking Bradley!"

...

"Shit. No one's coming out."

...

"Monty, get out and help. I'll cover you!"

...

Walt came back down a few seconds later wearing confused rage on his face because Monty hasn't gotten out yet.

"What the fuck!?"

Well Monty was at the back door, turned around and desperately struggled to get it open. We got tagged hard enough to bend the combat lock designed to keep the doors from being ripped open during a blast. But now we couldn't get the damned bolt released. It was stuck in a locked position. (Combat locks do this fairly often in IED strikes. It's happened twice to me.) We could have pried it open if we had time to bend an inch-thick bolt of steel by hand... we didn't. I turned around and saw Walt trying to help Monty so I yelled to him.

"Hey, hey, I'll take the gun! Just go out the fucking top. Go!"

Walt paused another nanosecond.

"Yup, that's good." And he was out. I took up the gun, Monty got back to work on the radio, and Clark hopped in the driver's seat. I turned toward the alley that we were supposed to be facing and took up my sector of fire (where my gun needs to be aimed). It also happened to be an angle that provided me with a crystal-clear view of the vehicle on fire behind me. This all happened within the first minute after the blast. I was scared, confused, and not entirely sure what was going on.

Walt was now at the Bradley with fire blanket and extinguisher in hand. Those in the trucks ahead figured out that there had been a hit and a raging fire had broken out on one of our security vehicles. I relayed information as fast as I could to the pertinent parties in my truck. Smoke had begun billowing and it was getting harder to see.

Guys from the front were now out of their trucks and running back to us with fire suppression and medic kits. The platoon sergeant and five or six Joes provided cover for extraction. We flooded the area with people pretty quick.

The fire had gotten so hot that all of the Bradley's armament had started cooking off. Cooking off is what happens when you apply enough heat to the primer of a bullet: the bullet casing explodes. The rounds weren't loaded in a gun to be aimed down a barrel, so bullets were firing in every direction. Thousands of rounds were now going off. We had at least six guys on the ground trying to help the boys in the Bradley. It was a nightmare shit show.

At about the time the others from the front got to Walt, the Bradley LT sprang from a hatch, coughing and screaming that Shaffer was still trapped inside.

With the help of everyone on the ground, Shaffer was eventually dragged out. It took several minutes with men burning their gloves and hands trying to wrench him out. That's several minutes of Shaffer burning. What I saw next I will never forget. They say the smell of burning flesh and the sight of a charred human is one of the grisliest sights to witness. This is a gross understatement. Shaffer was screaming wildly and still burning when he was finally pulled from the wreckage. A trail of skin and burned fatty tissue followed him like a paint smear. There was nothing left but a screaming figure, barely recognizable as a man. The Bradley driver did what he could with his extinguisher, but it was far from enough. Even though he'd been pulled out and there were no more flames, the smolder effect is something you don't think about. He basically had hot coals for flesh. He lay there on the ground whimpering and slowly burning. We used every extinguisher and fire blanket we had, but it was nowhere near enough. The lieutenant and driver of the Bradley were okay. The lieutenant had some smoke inhalation problems, and the driver had a piece of metal stuck in his neck. But they were basically fine. Shaffer was quickly loaded into one of the two remaining Bradleys and both CASEVACed back to Ramadi. CASualty EVACuation—think of an armed, armored ambulance with an escort.

Later we would be told that Shaffer suffered burns to 95% of his body. He might survive. But we couldn't focus on that. We had more work to do. Now added to our mission was the task of securing whatever sensitive items we could from the burning vehicle. Radios, maps, weapon systems, things like that. It was getting ugly. The fire had started to detonate some of the larger ordnance, the hundreds of 25-millimeter, high high-explosive rounds, and a few TOW anti-tank missiles. When you burn explosives sometimes they don't explode. Sometimes they just fizzle and burn really hot. It's pretty much 50/50 on what's going to fizzle and what's going to turn you into millions of tiny, squished bits spread over hundreds of feet. We were combat engineers, trained in explosives. We were well aware of the pretty big *Oh, fuck* moment we had on our hands. The boys grabbed what they could, and we cut our losses. We'd been in the same spot for way too long and people were now watching the chaos from alleyways and rooftops. Some were even cheering. The entire time this was going on, I'd been up in the turret less than twenty feet from the burning vehicle and all of its fun mini-blasts and bullet fire. Just doing my job. Avoiding shots from the cookoff fire, maintaining observation on the gathering crowds, updating my truck commander as the situation developed, and helping direct some of the guys on the ground (smoke was making it difficult to see for those closest). I wanted to come down; I wanted to close my eyes and hide. I was scared shitless every time a bullet hit my truck, a mini-blast went off, or one of those cheering the destruction fired an AK in celebration. But my piece of the pie was in that turret, so I stayed. Seeing everything and doing nothing.

Shaffer was CASEVACed within several minutes of the blast. Recovering gear, however, went on much longer, until we had to give up and let it burn. Eventually we did. After what felt like an eternity but was really only around an hour, the chaos was finally over. And when everyone started moving back to their trucks and I returned to my driver's seat, I sighed in both agony and relief. I was tired of being shot at by a mindless fire. I was tired of celebratory gunfire from the denizens of the neighborhood we were trying to help. I was tired of being hurt and angry. I was tired of being unable to shoot them for celebrating this horror show. The sights and smells were overpowering. I was so angry and afraid—I felt dozens of emotions all at once. My truck had no commo. Security was gone; they were the ones who had taken Shaffer back to base. The lead truck couldn't go in reverse, which isn't so bad, but you can't investigate any possible threats. You couldn't be wrong about whether your vehicle could fit through something or not. A lot of us had been through so much. Not just with Shaffer, the blast, the bodies from the tank round, and the cookoff fire. The celebratory gunfire itself was emotionally devastating. People just stood in the street dancing and a couple of them shot guns in the air because they had successfully hit us. There was nothing I could do. They weren't shooting at me, and I didn't have proof they were responsible for the IED, so I couldn't engage. Emotionally I was a wreck.

By now it was late in the dawn of the 14th. With no security, two broke-ass trucks, and not a lot of time left before sunup, we called the mission and decided to return to base. There were still bodies in the road and to avoid them, I cut it to the right over a big, giant curb.

Jostled the whole truck like crazy, but I managed to avoid the pile of parts. I did not, however, avoid a little dead boy. I hadn't seen him around the corner. He was up against the wall that I was cutting really tight into. My turn was going to put me almost right against the wall because I wanted to stay on the sidewalk as far away from the road as possible. By the time I saw him, it was either hit him or turn and hit them all. In a ten-ton vehicle, I didn't hear anything; didn't hear the bones crunch. But I knew. We didn't have rearview mirrors in the first-gen RGs.

I asked Walt, "I... I j-just saw that, uh... th- that kid at the last second. Did I—?"

He sorrowfully lowered his head and cut me off. "Yeah... just uh, just keep going man. You didn't do it on purpose."

With my heart now as broken as the boy's body, I drove forward and resumed my truck's position in formation. The new route took us down a highly used set of paved roads with a lot of traffic and overwatch positions. We selected it because the likelihood of IEDs was lower on roads civilians took often. We couldn't afford any more hits. Basically, we took the easiest route possible, and the entire platoon limped its way home in utter defeat. Less than an hour and we were back at the front gate. The drive was completely silent. The radio was still broken and no one in my truck said a single word. Just thick, heavy silence. If it had worked, I doubt the radio would've chirped at all.

When we pulled into base, I don't think we even fueled up. Most of us just parked, off-loaded gear, and went to bed. A few decided to wait up the extra half hour for morning chow to open. Not me; one job still needed doing. I waited until everyone got their shit out of my truck and headed over to the water buffalo in our motor pool. A water buffalo is just a really big water truck with a gravity-fed hose connected to the tank. As the sun came up on the morning of November 14, 2006, people were stirring and getting ready for their day. I was scraping the compacted remains of a child out of my tires with a spray hose and an army issue multi-tool.

When finished, I left my truck parked there and walked back to clear my head before meeting up with the boys for breakfast. It was about a ten-minute walk from the motor pool to the huts. Ten minutes in complete silence with nothing more mentally taxing than walking after the night I had just experienced was far too long. My brain reviewed everything that happened and finally registered the experience as reality. Truck swap, dead radio, grenade guy, bodies in the street, Bradley strike, fire, the cookoff, cheering locals, Shaffer, little dead boy. Over and over, my brain replayed each moment in every gory detail. Yet somehow it replayed at a blinding and crippling speed, and I couldn't stop it. The images and sounds were suffocating. It seemed as if there was no more air in all the world. Weakly, I leaned behind a tall concrete barrier and collapsed into the dirt. Tears fell from my face. I needed a moment and lay there for a good while. When I was able to collect myself enough to get up, something else dawned on me, which I had completely overlooked and forgotten. I wiped the dirt off of my weapon and uniform as best I could and muttered, "Humph... happy birthday me—big two-three today." And off I walked to breakfast.

There are a few obvious reasons why that night affected my life so much. But the ones not so obvious are what taught me to Always Try, *Semper Tentare*. One of them being that was the first night we had failed. We had been hit before, but that was pretty much in the job description. We weren't invincible but our vehicles were designed to take blasts, so it was an accepted risk. No one outside the team had been hit before. What we had missed in all the commotion over the bodies up front ripped right through the vehicle of the very people we were clearing for, supposedly protecting. Because of our failure, a man burned for almost four minutes. Shaffer did not die that night. But the first week of January, our platoon received an update notifying us that on December 27th, after a visit from his family, Shaffer succumbed to his wounds and died. Forty-four days after it happened, he lost his battle. I cannot express how horrifically he was wounded. In all my years with the army I have never seen someone injured that gravely who didn't die within hours or minutes. And I've seen more than my fair share. Make no mistake—I knew he was dead the moment I saw him pulled from the vehicle. There was no coming back from that level of trauma. He was just so damned tough he held on until he was able to at least be in the presence of his family one final time. That is the reason *Semper Tentare* became more than a phrase. If Shaffer could be that strong for forty-four days of unimaginable suffering, I could and would Always Try to ensure that it never happened again. No matter how hard or awful things were going to get, I would keep going until I had nothing left. Just like he did.

I had been on a deployment before. I had fired my gun and made things happen. Watching the pure hatred of celebrating a man who is burning to death and being powerless to do anything about it was new. It's hard to describe; war is endless boredom interrupted only by the worst aspects of humanity that occasionally intertwine with some of the bravest. I will never forget feeling totally and utterly powerless to save a man. My only job was to protect the guys on the ground, to make sure the people shooting their guns in celebration didn't decide to take up more lively targets. But they never did. I watched a brother in arms burn to death while people less than a stone's throw away celebrated it. I was helpless, couldn't even walk over and push them off. I saw that war is not clean and clear, even when it is. In film, violence and death is always succinct. Hell, guys that do take a long time to die always take just long enough to be cinematically advantageous. You know what they don't do? They don't linger long enough to be horrific. In real life, folks cling to survival for as long as they can. In real life, their deaths are gruesome and heartbreaking. When I think about it now, I don't cry profusely or smash everything in sight like I used to. But each time the memory of that night and I sit down for a visit, the inescapable odor of a truck and fully geared soldier burning, the joyful and torturous tandem screams that followed, the panic and fear, all of it comes back in vivid detail.

My naive belief that people really wouldn't do that to each other was destroyed. As a kid, I lived in southern California during the L.A. riots in '92. Still, the brutality of humankind didn't sink in until I saw it firsthand, in depth. To sum it up, the night of the Bradley was the first time I witnessed true, real-life horror with a front-row seat. On other occasions I'd been a participant, but never a shackled observer. That night broke me.

I didn't know it for years, but it did. It broke my belief that human beings would ever figure out how to stop killing strangers over shit that didn't matter. And, for all of you that have never witnessed violent, intentional killing and mayhem in real life, I am here to tell you movies cannot duplicate that. I have watched films and thought, *Damn, that's really close*, or *That's so accurate*. But the one thing a movie can never show you is what fighting in war does to your heart, soul, and mind. How it slams on you like a ten-ton hammer every day, all day, cracking and eventually shattering who you thought you were. That hammer broke a lot of what I believed were pillars of strength inside me. It's taken me (and a lot of other caring and patient people) years to fix them. Still rebuilding some of those pillars—always will be.

On the other hand, I would never trade away that night. From that experience I realized the kind of man I needed to be and hoped I could become. I learned about being a man of character and action by observing what went on that night.

I guess that's what made it both so horrible for me and so hard to face. I mean, shit, I've put people taller than me into bags smaller than me. Held friends with brains poking through broken pieces of skull. I've seen people disemboweled, cut in half, quartered, sometimes damaged so badly that nothing identifiable as human is left. Once, I shot the top of a man's head clean off—like the whole top. Nothing struck me as powerfully as that night. Like a mighty storm on the ocean disintegrating a rowboat, combat tore me to pieces. It wasn't the first time I'd had a bad mission and it wasn't the last. It was, however, the first time I witnessed so much inhumanity all at once. It was also the first time I felt responsible for the death of someone directly due to my failure. I know I don't bear sole responsibility for what happened that night, but enough of the weight is mine.

Initially it was too much. War had given me an intimate understanding of hatred and I wasn't strong enough to hold it at bay. My remaining time in Ramadi saw me morph into an angrier and more hateful version of myself. For the rest of the tour, I became everything I had witnessed. My small handful of kills were clean and justified, but the way I felt about them wasn't. I had become blood-drunk and felt intense joy taking enemy combatants' lives. To deal with those emotions and the shame that eventually came along with them, I drank thousands of dollars in alcohol my first few months back home.

My world collapsed into drunken decay and vitriolic outbursts of poorly aimed rage. Years were spent, off and on, destroying my own life. Also in my wake were numerous ruined personal relationships. I barely kept it together enough to do army functions a few days a month and hold down a civilian job (which, in truth, didn't last long). I hurt myself and I hurt others. Once, I even broke my own nose running completely through a gated fence that had just been shut in my face. I had been chasing someone in a drunken rage and they were running for their life. I earned it. But over the course of about a decade, with plenty of ugly days, worse nights, explosively emotional spiral rides, hospital visits both physical and mental in nature, that weight made me stronger. I grew with my burden, so to speak. It's funny because I'm sitting here trying to think of how to say:

Without the night of the Bradley, I don't know that I would have been prepared for leadership in Afghanistan. And without Afghanistan, I don't know that I would have ever gotten past the night of the Bradley.

But that's it. That night was the new cornerstone for who I would become. Years later, Afghanistan would complete the foundation. I used to view the night of the Bradley as something awful that happened to me. That was the wrong way to look at it. It still hurts and is awful to remember. But the memory no longer devastates me as it did. Instead it is now what has concreted my resolve to continually become better. Even in the face of loss or defeat I will pick myself up, deal with the aftermath of failing, and keep trying. Always.

Chapter 3

Bull

I was once thrown to the ground and screamed at by a guy named Bull. His legal title was Sergeant Kortney Reese Jensen, but everyone just called him Bull. He was the skinniest redneck I ever met. First time I saw him was his first day at the unit, sometime in 2003, I think. He was wearing a beat up, ten-gallon hat atop his shiny, bald head, a plain white tee, jeans too tight to move in, a worn leather belt held up with a big, shiny cowboy buckle, and some dirty-ass brown boots. He worked odd jobs before the army, mostly labor intensive. Bull's memorable physical and character qualities were great in number, but for me, he was defined by his hands. For such a tall, skinny fella he had these thick, scarred, calloused, powerful hands.

Over the course of our friendship, I was on the receiving end of a few of his punches. His hands were wrecking balls. Overseas, there were many times we would be working long hours putting our truck back together or just generally fixing some of our broke shit and damn if my hands didn't hurt. They'd shake and crack, even bleed sometimes. But I always kept working because if I didn't, this skinny hillbilly would effortlessly put me to shame. He never bragged about anything either. If he finished up work before me, he'd pop over and help out so we could both go get some chow and call it a day. He just liked people and lived for helping out. Don't get me wrong, he didn't crap sunshine or piss rainbows. Bull was as tough as nails, a moderately short-tempered fighting man of honor if ever there lived one. He just always carried himself in a manner that made everyone like and respect him, even those he occasionally pummeled.

In April 2007, Bull and I were on our second deployment together. This time to Ramadi, Iraq, and the 26th was a maintenance day for our platoon. Someone had put the wrong type of fuel into a couple trucks, so my task for the day was replacing ruined fuel filters. Meanwhile, 3rd Platoon had a big mission going on in Albu Bali, we called it "the sticks." It was a province of open land used mainly for farming, with some very large plantation-style properties, a lot of rudimentary shacks, and one enemy bomb-making facility that needed removal. Third Platoon had been hit hard and needed a new lead truck with a crew to man it. Our platoon boss, First Lieutenant Peterson, or Lieutenant Pete was ordered to pick four soldiers to outfit a vehicle and get out to Albu Bali, asap. Our platoon had been through a few difficulties of our own recently and really needed this maintenance day. In the first four months of the tour our platoon had already replaced three squad leaders due to blast injuries. Bull was one of the guys promoted into the newly vacant slots and was now in charge of Third Squad, which I luckily happened to be in. In the end, Lieutenant Pete told Bull we had a mission and to grab two fresh bodies. Bull was with Corporal James Marshall (Marsh), and they caught up with me walking over to the motor pool. Bull waved me down.

"Hey Mayfield, Third got hit pretty hard out in the sticks. They need a truck and crew. You really wanna fix someone else's fuck up," pointing at the new filters in my hand, "or you wanna go save the day?"

A big thank-you smile washed over my face. "Hell yeah, give me five minutes. I'll be at the truck with gear and water." And I took off running back to the hooches for my armor.

With prior duties properly ditched, we had the truck spun up with weapons, water, gas, and ammo in fifteen minutes. We had Lieutenant Pete up in the gunner's turret; Marsh was extra eyes up front; Jensen was in back running the radio and talking to each landowner as we passed through their sectors; and I was driving. We headed on over to the gate and since a minimum of three armed and armored vehicles were required to leave Camp Ramadi, we were given an escort of three Humvees. They were an experienced emergency response team, so our link-up only took three minutes. Once we were all talking on the same frequency, we rolled. I had never, and would never again, drive that fast down Route Mobile. About 90 kph was the max our vehicle could go. Mobile was an actual highway with big lanes, so the first part of our journey went quick and easy. After twenty minutes, we hit the specified GPS marker and turned off the paved road into dirt and moon dust territory, leaving our Humvee escort to turn around and return to base. After another twenty minutes of much slower and moderately more cautious travel, we arrived at Warrior Strongpoint, a combat outpost under US control serving as the headquarters of the troops stationed in Albu Bali.

Warrior Strongpoint was one of those nicer, plantation-style estates. It was made of white-painted exterior concrete walls, the inside was a lighter blue I think, with granite floors throughout and a huge courtyard. The entire property was surrounded by a high, concrete wall with only one space large enough for a drive gate. It served quite well as a small outpost and instead of a gate, the open space was closed off with a manned Abrams Tank. If you needed in, the Abrams had to let you in. The courtyard was littered with people, trucks, and equipment. Inside the building was a makeshift command center, hospital (first aid station stretched thinner than Gumby), and soldier housing. It was a grand and spacious building when considered a family home. As a base for tactical operations it was functional, but beyond cramped. The path leading to Warrior Strongpoint was well-trod by military vehicles and they knew we were coming. The Abrams was in reverse, moving as we pulled up. We got inside and immediately attended a brief of the situation. A massive waiting game had begun. There was one road leading to the bomb factory from two directions, Route Dusters. Second Platoon would clear Dusters' North, while Third Platoon took care of Dusters' South. Everyone would meet in the middle near the factory so ground troops could roll in, get staged, and begin the assault. While Third had been down for only about an hour, they had progressed farther than Second. Second Platoon had several IED finds and an enemy engagement with RPGs slowing their progress. Higher up consensus was to hold off getting Third Platoon back into action until Second Platoon's progress had caught up. Nobody wanted half our forces stuck at the front gate of a well-manned and well-armed bomb factory, waiting for the other half to get there. Second continued forward for nearly five hours before we got rolling. During those hours, they endured an IED strike, a few shoot-outs, another RPG attack, and two IED finds. My time waiting out those hours went differently.

After the first hour inside Warrior Strongpoint's walls crawled by, we donned our armor and got back in the truck. We wanted to be ready to go the second the order dropped. After all, it was only around 10:30 and a lot could still happen. At first, we game-planned, going over our established tactics and procedures, double-checking gear, and rehashing the mission's details so we knew everything backwards and forwards. Our piece of the mission was simple enough, though, and we eventually ended up talking about life, how crazy our tour had been, and all the things we were going to do when we got home. Looking back, that span of a few hours tucked in a truck with three brothers were some of the best of my life. It was a warm, beautiful day with only the tiniest, white puffy clouds barely dotting the skyline. The occasional Apache helicopter would fly overhead and I remember just listening to the hopes and dreams of my friends, while I watched a giant, metal bee buzz around the sun.

Sometime around 1500, we got the order to roll onto Dusters and begin clearing up to the rest of Third Platoon. After waving thanks to the Abrams boys, we took a right out the gate and headed up onto Dusters. Dusters was a raised dirt road, maybe twenty-five feet wide in the widest areas and maybe only ten feet in the tight spots. It was essentially a large and long dirt berm to aid with irrigation that had a dirt road established on top of it. To the right of the road were fields for farming, on the left were small villages separated by more fields and natural tree groves. There were Bradley Fighting Vehicles positioned on top of Dusters every few hundred yards to keep overwatch of the portion of the route previously cleared by Third Platoon. We must have passed just short of a dozen before getting all the way up to 3rd.

After rounding a bend covered by a grove of palm trees, Third's rear truck came into view. Bull had been on the radio with Master Sergeant Dave Johnson (or BFD for Big Fucking Dave). We were going to slow-roll past everyone. They would just fall into the order of march with us and continue mission. Everything went smoothly for a few kilometers —just slow and methodical clearing. We came upon a blast hole from our Husky operator Specialist Rodney Mitchell's hit a few weeks prior and did not feel good about it. We drove right up, looked in the blast hole and talked it over in the truck, then on the radio with the rest of the platoon. BFD was also not satisfied and had the Husky come forward and roll over it. He got no hits on his metal detector panels. We then shuffled the Husky back and the Buffalo crew pulled up to use the interrogating arm. They poked around for twenty minutes and couldn't unearth anything suspicious, so they reversed back into formation. Once the vehicle shuffling was complete, the hole was deemed clear. We pulled up to the lead again to resume clearing, for two feet.

We moved to the left of Mitchell's blast hole and were immediately catapulted into the air by a large detonation. I was knocked unconscious. This next bit is all completely the word of Master Sergeant Johnson in the truck behind us. He said as soon as we passed Mitchell's blast hole, the ground erupted and put the back end of my truck six feet in the air. We were hit in the front, which went much higher. Master Sergeant Johnson said it looked as if we would have gone end-over-end had it not been for an additional 750-lb. gyro cam mounted to the left-rear side of the truck. The added weight kept the truck down and to the left, preventing it from completely flipping. An IED had been placed underneath the road using the side of the blast hole as the entry point to disguise the fresh digging as old blast debris.

I woke to a brown, screaming world. There was dust everywhere, but my vision had gone kind of putrid yellow and I couldn't see more than a few inches from my face. People were moving but there was no sound, except for a metal banshee wailing in my head. Bull was checking on Pete; he was hurt. Marsh was still out but didn't look too bad.

Everything went dark again.

I was sort of on my feet and Bull was throwing me out of the back of the truck. BFD was there with a mad look on his face to catch me. Everything was still miscolored and there was no sound but the banshee. The sky was pretty, kind of green. I didn't know what was happening. I started wandering around to get my bearings.

Bull's full, armored weight slammed into my body, violently knocking us both to the ground in a pile of camouflage, blood, and dirt. An RPG zoomed past where I had just been standing and exploded faintly off in the distance, my hearing coming back but still muffled. Bull, still lying on top of me, was screaming something. He wasn't as loud as the rocket, so I didn't hear a word and just smiled at him, confused. I'm certain my behavior would have been more useful if not for a small piece of shrapnel lodged firmly in my forehead.

We were loaded into BFD's vehicle, a CASEVAC truck, and taken back to Warrior Strongpoint for medical. Turned out all four of us were in pretty bad shape. Lieutenant Pete had hit his ass on the unpadded metal seat and crushed his tailbone. Jensen hit his head on the ceiling, causing what we later found out to be a compressed C-spine and some brain damage. Marsh broke through his safety harness during the blast and hit pretty much everything, getting really banged up. I managed to break a few bones and somehow grow a small, metal unicorn horn. All in all, Pete was the only one of us who was able to stay in country. Bull lied his way out of medical evacuation initially but was later sent home when he deteriorated mentally instead of getting better. Marsh and I were sent to Logistic Support Area Anaconda, a much better-equipped medical facility and, at the time, the biggest US-run base in Iraq. This hit was what led to Bull's and Marsh's medical retirements. I would be in a hospital for a while but, since I never liked those and wanted to return to my brothers, I left on my own as soon as I physically could. I'd been there a few days and had been asking staff constantly about when I was going to be able to rejoin my unit. Must've cooked a few nerves with the staff. Not sure if it was to shut me up or what, but someone told me that I was going to be returning to my unit soon. They were just waiting on paperwork for travel. That was a half lie. They were waiting on paperwork to approve my transfer to a hospital in the states. Hours passed with no new information. At completely variable intervals, I checked in and would be frustratingly told something like, "Not now. We'll get you when it's time."

Helicopters flew in and out of Anaconda's hospital pad nearly nonstop. When I felt particularly ready to not be in a hospital anymore, I donned my uniform and walked to the helipad to wait for an outgoing bird heading to Ramadi. It didn't take long. After lying about already having my Return to Duty status and waving my medical papers far enough away so the pilot couldn't read them, I found my ride. The flight was quick and without incident. Since my copy of the damning medical file was no longer useful, I tossed it into a burn barrel on my way back to the hooches and told command it had been misplaced during transit. Accepted as truth, my lie got me back to work fairly quickly. Gone only about a week and I had already made it back. However, my subterfuge fooled precisely no one. After a few weeks of piss-poor soldiering and some very concerned observation, command ordered me back to medical for re-evaluation. I was immediately sent to Landstuhl Regional Medical Center in Germany, then on home to the States for continued care. In truth, all four of us should have gone back to the States that day and we knew it. We just wouldn't leave everyone as long as the choice remained ours to make.

Months later in September of 2007, while I was in Madigan Army Hospital near Seattle, Washington, Bull called me from Brooke Army Medical Center in Houston, Texas. He was being released to go home and was out-processing at Fort Lewis, where Madigan is located. A couple of days later we were having pizza in the common room of his temporary barracks. He seemed so much more "there" and excited to go home to see his wife and kid. We caught up on who was in which hospital and how their recoveries were going. Our tour was over but, by now, so was the rest of our company's. For the first time, we could laugh and talk about things without feeling guilty about not fighting alongside our brothers. They were home now. We argued a little about my being released back to duty. Bull was convinced I should have been retired like him, but the joy of going home kept the mood light.

We stayed in touch, seeing each other frequently as we settled into our newly adjusted lives. I went back to the civilian world and resumed my job. In the reserves, I was promoted and would be in charge of my own squad from now on. Bull medically retired from service and became a full-time family man, and he was good at it. We often had cookouts or reasonless visits to each other's houses. One day in his backyard, I was trying to understand how to instill bravery or inspire my guys the way Bull's actions inspired me.

So over a hot-dog-grilling session, I asked him, "How come you always stepped right in to save my ass? No hesitation, ya know, no thoughts of yourself." Amused by my question, he set down the tongs, put on his patented, reassuring smile and plainly stated:

"Because you're worth more."

He squeezed my neck with his vice-like hand, retrieved his tongs, and resumed his dog-flipping. That was the end of that. I didn't understand his answer until I had a squad of my own in live combat. It wasn't that Bull felt he had less value, it's that he chose to uplift others' lives above his own. He did this for me in particular because he knew I did the same for him. We'd been through a lot together by the time the RPG debacle occurred. Among others, he was on the ground the night of the Bradley fire. By then he was family to me. When he said I was worth more it was because I was one of his protectors. If you are surrounded by a platoon of soldiers who each put their personal safety dead last, you never have to worry about yourself. I certainly didn't. All I had to worry about was doing my job well and sometimes doing the ugly stuff, so someone else could take a breather.

A couple of not-so-rose-colored years passed, and life seemed to establish a kind of normalcy. It was now July 2010. Bull was a pillar of his community, had been promoted in his motorcycle club, and had another baby girl on the way. I was at Fort Irwin National Training Center in California, gearing up to deploy to Afghanistan. Right before that, Bull and I went to lunch, and I told him about trying to get medically cleared to deploy again. I had been returned to duty, sure, but I wasn't cleared to be a fighting soldier. I was still barred from deploying to a combat zone. The waiver to be cleared was already sent up, but I had dragged my feet telling Bull because it wasn't even a sure thing yet. Mostly I hadn't told him because I knew what would happen.

He immediately exploded in fearful anger at me. With a tone of rage, sorrow, and fear, he detailed the concerns he had. We both knew I couldn't take too many more hits before something bad permanently happened to me like it did him. He told me how he hated that he wasn't a soldier anymore, that he felt he didn't know his place now. If I did go, who would beat my ass when I was in need of an attitude adjustment? Who would be my backup in both bar and firefights? Who would force me to watch old westerns with him while I cleaned my weapon and sharpened my knife? Who was going to get me into, and back out of, trouble? Who was going to fail miserably at out-drinking me? Who was going to throw me to the ground and save my life? According to him, those jobs were his and he had trouble trusting others to do them. The conversation was long and not easy, but we agreed that when I got back from National Training Center, we'd go camping and talk this horseshit out for real. We talked out in the middle of nowhere with bottles of whiskey and our fists.

National Training Center is located in Death Valley, California, in the Mojave Desert. My whole company of over a hundred-and-twenty souls lived in one giant tent for a few weeks of route clearance training in the place with the highest recorded air temperature on earth. It was damned hot. The tent was empty except for a large pile of folding tables for our makeshift command center and enough cots for everyone. We ran missions in the morning, indoor class training in the afternoons to avoid the heat, and squad tactics in the evenings. On the morning of July 31st, my platoon finished the morning training mission and was conducting the After Action Review. This is where the platoon and trainers all gather together in a circle and discuss how the operation went. The good, the bad, and the *What the fuck were you thinking?* moments all get hashed out to improve platoon cohesiveness and efficiency.

While we were all in our huddle, the company commander, Captain Scott Sparrow, came over a hill and drove right up to us in the duty van. This was strange because we weren't allowed to have nontactical vehicles on the training lanes. A non-armored, fifteen-man passenger van was definitely atypical out on the lanes, but for our company commander to stop an After Action Review was unheard of. He would come and listen in sometimes, even inject his own thoughts from time to time, but never halt one. He also wasn't angry, but something was clearly wrong. He told Brummett and me to come with him immediately. Brummett handed over platoon sergeant duties to Jones. I handed my squad off to Vargas. We got into the van and away we went. We asked Cap what was going on, but he was solemn and wouldn't tell us anything. We rode in confused silence for a few minutes until we pulled up to some trailer classrooms and walked inside. Cookie, Pohle, and Lissy were there, waiting with Young. They all looked like shit, but I was still glad to see them. Everyone in that classroom had served together in Ramadi. Young had been promoted to platoon sergeant for Second Platoon, so it made sense he was there. But the other three were not part of the 744; they were all engineers in the 455 now. I didn't even know their company was at National Training Center, but it didn't matter. I loved these guys and was overjoyed to see them.

A huge smile erupted on my face, and I threw my arms around them one by one. Awkwardly, they weren't that enthused to see me. I didn't understand. Captain Sparrow broke up the one-sided glee fest and asked me to take a seat when everyone else in the trailer was still standing. Suspicious, I refused and had honestly had enough of the cryptic shit.

"Sir, I'm not sitting. What the hell is going on?"

Young leaned into Brummett and whispered something. I couldn't hear but his face went white as newly fallen snow and his eyes welled up. He put his hand on my shoulder and nodded shakily.

"Dude, sit down."

At that point I was furious. I didn't need kid gloves. I figured I knew what happened—word had come down and my deployment waiver had been rejected. Maybe the guys from the Ramadi tour were there to soften the blow or whatever; I didn't need coddling.

I was getting out of hand fast, so Captain Sparrow relented to let me stand as he sorrowfully informed me that Bull had died in the late hours of the night and that initial reports suggested it was suicide. I should have sat down. My brain struggled to process what it had just heard. I stood in silence for a few seconds before slumping to the floor as if all my bones had somehow instantly disintegrated. Even now, more than ten years later, tears are forming as I type this. I was inconsolable for the better part of an hour. Brummett and Sparrow talked together and decided to release me from duty for the remainder of the day as I was, off and on, a useless, babbling, tear-filled mess of a man. I would compose myself for a moment and then another wave of loss would crash over me again. I walked deliriously all over the boiling wasteland that is National Training Center trying to come to grips with the possibility that someone who had lived so bravely and selflessly could be guilty of suicide. Not possible. There had to be something else. Something nefarious. People like Bull were the stuff of legend. Hell, he *was* a legend. Those kinds of folks don't fall to suicide. Right? But he was dead, that was a certainty. I had to know the truth. I called McCarran International Airport to see about tickets back to Salt Lake. Nothing would stop me; I was going home to say goodbye to my best friend and get some damned answers. I got pricing, dates, and times for several different flights, as I wasn't sure when the funeral would be. But I wanted to be as prepared as possible when I approached Captain Sparrow about letting me go home for the funeral even though we were still in training. I had to go. It was my fault he was dead.

At least back then I believed it was. The reason I blamed myself was a missed call earlier that morning. About half an hour before he died, Bull called my cell phone. In a tent of so many sleeping soldiers, a phone blaring in the middle of the night was extremely unwelcome. I scrambled to get to it, saw it was him and figured he was drunk and wanting to plan our camping trip. Since about thirty people were already hollering various versions of "Turn that shit off," I silenced my phone and resolved to just return his call when I had a free minute. I ignored his call. I could not ignore his farewell.

By the time I returned to the tents, it was late evening. Captain Sparrow had already acquired the funeral date, organized a four-day pass to allow me to fly home, and arranged transportation to the airport. All I had to do was thank him and book the flight. The 455th would also be sending Pohle to represent their company. The duty driver would take us to the airport on Wednesday and pick us back up on Saturday. If you're wondering why the 455th also sent a guy, it's because the 455th used to be called Bravo 321st. Bull was well-loved and respected by hundreds of soldiers.

On August 4th, Pohle and I took a flight back to Salt Lake. My parents met us at the airport with my truck. After thanking them, we drove directly to Ogden to stay at Bull's older sister Brooke's house. She and her husband opened their home to a total of fifteen of us who didn't live close by. They fed us and let us use their home for a three-day live-in wake. They were more gracious than any hosts should have been in light of their loss. More booze was brought into their home than you'd find at a liquor store, and we drank all of it.

That first night we stayed up late, telling stories and pounding bravery fluid. The sun woke up before the last of us went to sleep. We'd reminisced on all the antics and craziness we had been through together with Bull. I told one of my favorites. There was a time on our first deployment when our platoon sergeant, Sergeant First Class Decloute, came looking for Bull to fill out some paperwork. We were between paving missions, and it was a slow afternoon. I was lounging in a chair in the hallway that linked our platoon's living quarters together. It was close to chow time, so I was sharpening my pocketknife to eat up the last few minutes of waiting. Well, the platoon sergeant saw me and made a moderately aggravated charge right on over. She had a bunch of papers in her hand and was waving them at me.

In a not-so-happy tone she inquired, "Mayfield, where is Specialist-who-wants-to-be-promoted-to-Sergeant-Jensen? He needs to have signed this damned promotion packet *yesterday*."

I knew where Bull was but wasn't telling. In plain view out back through a glass door about fifty feet down the hallway was Bull, a gas can, and a beetle. That visual kind of reminded me of those 1950s-era-style posters where some mischievous child is about to do something really stupid and learn a life lesson. You know, "This is Timmy with a bottle rocket and cattail. What could go wrong kids?" At least that's the image I got.

"No, Sergeant D., haven't seen him. Any specific message for him when I do?"

I discreetly glanced at the door to see if he was still there. At first, all I saw was a flaming gas can go flying through the picture frame of the glass door, no visual on Bull. A few seconds of mild bewilderment later, he popped back into frame, hopping and yipping around like a madman because he was on fire.

"Tell him if he wants to make the board this round of promotions, he needs to come see me and sign this packet by close of business today."

Out of boredom, Bull had been trying to burn a beetle out back behind our hooches and it wasn't working. He grabbed a can of fuel and was going to pour it on the beetle but instead dropped the can and immediately set his legs ablaze. He had a lit cigarette in his hand as he fumbled to catch the falling, uncapped can.

"Roger, Sergeant. I'll let him know the moment he returns to cool his heels," was all I could chuckle out. It took all my concentration not to burst out laughing while he kicked up dust and flopped around doing the worst "stop, drop, and roll" I'd ever seen. He was in no real life-threatening danger and his leg was only mildly red and hairless when it was over. Damn-near peed my pants laughing the second my platoon sergeant took off to look for him elsewhere though.

Bull saw me laughing as he walked through the glass door, and with a huge, wide-eyed grin on his face he asked, "Did you see that shit? Fucking beetle out there is the only thing not on fire... today was not his day. Heh, heh, heh... was that Sergeant D. in here?"

I put my knife away and explained his signature was needed and that our platoon sergeant was none too happy having to hunt him down. "Alright, let's go eat and then we'll go to headquarters after." Didn't even bother to change his pants, just marched off to chow with a big burn hole where his cargo pocket should have been. But that was Bull, never embarrassed and nothing ever got to him.

The day of the funeral came, and just getting parked was nigh impossible. Hundreds of people showed up. There were maybe seventy-five parking spaces and most of those were taken up by cars. Each remaining empty stall was filled with two or three motorcycles. It still wasn't enough. The sidewalks, grass, and road leading to the funeral site were all lined with bikes. Pohle and I were in uniform as we had been asked to carry Bull's casket to the grave site. But there were dozens more in full dress attending. His seven-year-old daughter, Autumn, sang "You Are My Sunshine," and to hear the weeping that accompanied her sounded as if the whole world had broken into tears. Seeing that many people from so many different walks of life affected by Bull's death was humbling. Being able to attend his funeral and see his impact on the other areas of his life I wasn't part of only highlighted how big a loss his death was. As the funeral came to a close, Pohle and I said our goodbyes and off to the airport we drove.

On the flight back to National Training Center I couldn't get Bull out of my mind. All we had gone through together, all we had survived, and he succumbed to something as intangible as thought. My own thoughts kept visually returning to the sea of mourning, those in attendance at his funeral. The sheer number of people shedding tears for him in one space was crushing to witness. I saw how one man's choice left giant blast holes in so many hearts and once again, had to thank my friend. Even though the last decision Bull made in this world was awful and blindly misguided, he gave me one last lesson. For more than 99% of his life, Bull was content to simply give without taking. His final moment did not diminish that. It did, however, mark the first moment of horrible and undeserved pain for everyone who knew and loved him.

In the coming months and years, suicide would sing to me from time to time, and it was memories of losing Bull and all the devastated lives in his wake that silenced those songs. I would never—and could never—do that to the people I care for simply because I now understood how much suicide hurts. One of my best friends in high school shortly after we graduated, one of my uncles before I was born, and a large handful of vets I served with have all killed themselves. So it wasn't a foreign concept when Bull did it. But I idolized him. Not because he was good at sports or could talk to women, but because he saved lives. He was the example, and not just for me, of what unbreakable spirit truly meant. I had no idea he had been so torn asunder not being able to be that man anymore. Soon enough I would understand that pain firsthand. He's been saving me ever since. Every time my life has felt overwhelming and that it is simply too much to continue, Bull comes 'round and reminds me: *This pain is mine, I can never give it away, never push it onto another. However far it may be, I can and will carry my lessons all the way to the end.*

When I do finally breathe my last breath, I'll find Bull in the afterlife, beat seven shades of shit out of him, scrape him off the floor, hug him for a century, then grab a few beers and catch up.

Sergeant Kortney Reese Jensen, you dumb bastard. I love you brother.

See you in a long time.

Chapter 4

To Be Understood, You Must Explain Yourself

Getting medically cleared to deploy to Afghanistan was an endeavor that took the better part of a year to accomplish. Three separate packets for release were sent up to Big Army and all three requests were denied. The first was kicked back because it was filled out improperly. The second packet was denied outright due to my history of blast exposure and came with a nicely worded rebuke from a medical colonel. The email said things like "commendable to want to continue serving," "irresponsible to waste resources to clear med board with injury history," "other ways to serve," "Non-Deployable," etc. The third and final denial came in the form of a text from my First Sergeant the morning I graduated the Engineer Explosive Ordnance Clearance Agent School, or EOCA (eeh-oh-kuh), at Fort Leonard Wood in Missouri.

It was early September 2010 and three of us had been selected to attend this iteration—Ballantyne, Hansen, and myself. As the name implies, EOCA qualifies a combat engineer as Explosive Ordnance Disposal (EOD) tech level one, sort of. Oftentimes on mission in-country, we would find an IED and have to wait up to several hours while EOD made their way out to us to dispose of it. EOD techs were much lower in number than the demand for them. Big Army's easy fix was to create a school that twelve Bravos could attend and get certified to safely render inert and dispose of small-scale IEDs. Passing this course allowed me and the other graduates to simply blow IEDs we found instead of holding up a mission waiting hours for EOD to take care of it. This was a hard course with a high rate of failure. It was six weeks of memorizing how to identify and dismantle initiators, fuses, propellants, igniters, chemical munitions, marking systems on thousands of different mines, bombs, grenades, and rockets from different countries. It was a manic, sleepless, study all the time, tape your eyes open kind of course. After a grueling course, two out of the three of us in our iteration had passed. It was the morning of graduation, and I was feeling strangely optimistic about being cleared. This course alone was proof my mental capability was up to scratch. It was about that moment I received First Sergeant's text.

"Your shit's fucked up again. Call me."

I had been holding my EOCA 1059 for less than five minutes. A 1059 is a piece of paper that acts as the military version of a diploma. I actually passed EOCA and couldn't even enjoy that little victory. I lost my cool completely and blew up—full on mantrum. A mantrum is exactly what you think it is, an adult tantrum. You've all seen or had one. Someone gets so mad that a few things physically begin to change in them. One eye, occasionally both, becomes huge and manic. Picture a mad scientist without the scientist part. Foot-stomping and fist-waving soon replace fine motor skills. Finally a Cro-Magnon-like posture develops and loads of expletives and filler words that don't make any sense pour out of their mouth at an unbelievable rate of speed.

"These rat-turd-sandwich-pencil-pushing-shit-stain-banana-dick-asshole-tire-iron-beefcake-know-it-all-rainbow-fucking-sherbet-RAAAWWWRRR!"

I was standing outside one of the classrooms leaning on a desk that had been set up to evaluate bomb bot operations. Yes, we played with, and were now certified on, the bomb disposal robots. Only able to see red at that exact moment, however, I picked up one of the two desks, and used it to destroy the other. In a short while there was only me, some kindling, and a few metal desk legs remaining. One of the course instructors came out after I had finished.

"Hey, Sergeant Mayfield, you done breaking my shit yet?"

Slightly remorseful but more embarrassed about the desk obliteration, I spun around and stood at ease. This is a semi-formal position of respect when speaking to a non-commissioned officer of higher rank than you. It was also my immediate physical apology.

"Yeah, Sergeant First Class Greethum, I'm sorry about that. Got word I was pulled from my deployment and I kind of lost it for a second."

He was a big, red-haired, freckle-faced man with his years of experience deeply carved around his eyes like wisdom wrinkles. His grimace broke and he began chuckling warmly at me.

"Oh, I'm fully aware. You've been clearly and loudly wailing about it for the last three minutes. Looking at my desks, I'd say you're plenty capable of going. As you know, there is more than one way to get something done in this army. You passed this course, so you're obviously a smart enough guy. You'll figure it out. Until then, stop trashing my training aids and go away. You're done here. Oh, and good luck." He shook my hand and walked back inside without another word.

Ballantyne came over and helped me clean up the debris from my mantrum, then we left for the airport. He drove while I frantically called everybody I could think of. It was a panic scramble to say the least but after an hour or two chewing off a dozen different ears, there was a plan: meet with the state surgeon of Utah, Colonel Peter Taillac. My mom worked in his office, so I had an in. While he didn't have the power to just arbitrarily make changes to my profile and approve my wish to deploy, a letter from the state surgeon holds some serious weight. Before he would give me anything though, there were tests to determine my physical and mental capabilities.

Could I actually perform my assigned duties? Or was I really as broken and incapable as Big Army said?

I did the medical dance in my underwear, picking things up,

putting them down, touching my nose a bunch of times while varying which foot was on the ground, etc. After about an hour of reflex, motor function, and cognitive processing tests, he'd collected sufficient proof that I was good to go, sort of. Colonel Taillac explained that while I was clearly able to do everything required of me, there was evidence that damaging my brain any further would not be a difficult thing to do.

"I can't find anything that would bar you from being able to deploy. But son, a box, even empty, hits your head it will be a problem. I'll put a letter together to show the deployment doctors my findings. But they'll decide. Best I can do without breaking rules."

So, letter in hand, I was ready to tackle the next big issue with being removed from the battle roster. I needed to find a way to get sent to Fort McCoy, Wisconsin, where the main base was.

The base was a mobilization station, or MOB, where Reserve or National Guard companies are trained for deployments. Several bases across the US have this designation. I had already deployed out of Fort Lewis, Washington and Fort Bliss, Texas. Even though locations changed, the routine had not. Units live, eat, train, and sleep together for the first time and will continue through the end of the deployment. Bonds are formed and those in positions unsuited to them are moved elsewhere. Night missions, day missions, equipment training, fitness regimens, and death by PowerPoint brief nearly every day. Activities themselves are manageable. It's the hours that become occasionally grueling. To ensure deploying soldiers are as focused on training as possible, a small team of non-deploying soldiers called White Cell from the unit trains alongside them. Duties of White Cell include arranging barracks to stay in, scheduling training lanes and coordinating their use with base personnel, mail room, organizing transportation to and from training sites, ordering and withdrawing weapons and ammunition, duty drivers, and a slew of office work. After brainstorming with company leadership, my name was slapped onto the White Cell roster, and I flew to Fort McCoy to meet up with the rest of the company.

Aside from eight people, nobody knew I was on White Cell instead of MOB orders. I still trained as First Platoon's Third Squad leader. Between training, I made phone calls and set up a meeting with the MOB doctor. MOB doctors are the overseeing medical professionals at the end of the line. No matter what you do, who writes your fancy letter, how many officers stick their necks out, or how many tests you pass with flying colors, the MOB doc is the lynchpin. If they said yes, it was all worth it. If they said no, there would be no more attempts. But I was confident it was going to go well. My performance during training lanes was pretty good and I was determined to use that and passing EOCA as part of my defense. My plan in motion, the appointment for my pass/fail moment was scheduled. I was a worthless, nervous mess for about a week. I worked as hard as I could morning, noon, and night to prove my worth. But mostly it was just to keep my mind from wandering into *What if the MOB doc says no?* territory. Lissy, one of the eight who knew my plight, was constantly trying to keep my mind occupied. This was it. It was Friday night, and Monday morning I would have a real, no-turning-back-now answer. More than six months of fighting tooth and nail had boiled down to just one weekend of agonized clock-watching.

The meeting was fast and went beyond well. I forgot his name, but the MOB doc was already familiar with my giant file by the time I walked in the door at 08:30 that morning. As I stepped through the office door, he spoke up without really greeting or looking up at me. He had a blue medical file held up squarely in front of his face.

"So you're my problem child?"

Taken aback, I replied, "Um... I guess? Sir."

He slapped the fat, heavy folder onto his desk.

"Because of you, this bit of boredom was my reading material this weekend." He seemed angry and put off.

"Sir I apologize if I ruined your weeke—"

"You'll get a chance to speak uninterrupted. For now, shut up and let me finish. You aren't my only case today. But hopefully you'll be my easiest. I read this folder cover to cover, read all the letters for why you should or should not deploy. I have mostly made up my mind. All that's left is for you to put in your two cents and see which direction it tips my decision... Why should I let you go?"

Butterflies started doing acrobatics in my guts. "Sir, were you able to see my test scores from—" Interrupted again.

"Yes, I saw that. I don't care. Why should I let you go? Your words." He saw I was getting shaky and frustrated. "Breathe and take your time. You have a good case here in this folder. But a folder isn't a soldier." A third time and with more compassion than impatience, he asked, "Now, why should I let you go?" Thirty-seven-million years silently passed in that office before I opened my mouth to reply.

"Because I have to, sir. I've trained for three years, side by side, with these men. I gave my word. If I leave them now, how could I live with myself if any of them don't come home? I *was* hurt and I know that somehow makes me a bigger risk. I sort of understand that, truly do. But I'm really good at my job. I don't have children. I don't have a wife. I don't even have a fucking dog... Sir, I have soldiers. They have me. And yes, I may end up in real bad shape. Or dead." My head and hands began to vibrate, and my voice went up a few decibels. "But how is that any different than my guys, my soldiers who have no previous injuries at all? Experience. I have some. They don't. A bullet to the head would kill my replacement just as easily as it would kill me. Except I've already proved that I can do this job. Don't take it away from me because I *could* get hurt in the future. Every one of us could randomly die any given morning taking a shit. But we don't fire folks for what *may* happen. If I do get hurt again and can't do the job, then you can send me packing. Don't shelf me if I'm still useful sir. Please."

When I paused for his answer and took a breath, I noticed his eyes were considerably wider than a minute earlier. "Good God kid, you aren't intense at all, are you? You must have annoyed a lot of people to get here. As I understand it you are currently on White Cell orders, correct?"

Oh shit, here it comes, I thought and nodded guilty.

"I commend you. That's some pretty tricky business. Nothing I can do about that for a few days, but you'll get your deployment orders cut within the week. I am doing this not because your folder says you are capable, but because I honestly think there is no one capable of stopping you, except you." Victory tears began to form in my eyes. "I know you fought long and hard to be approved. Prove it to be the right call."

I vigorously shook the man's hand until it damn near fell off.

"I won't. Er... I will! You, uh, thank you so much, Sir. I can't tell you how much this means."

He wrenched his hand away and wiggled it around as if something was loose or busted. "Clearly. Get out and send the next soldier in. Make sure you get copies of that approval form to your unit and MOB personnel immediately." I sheepishly thanked him once more, hands free, and zoomed out the door with my golden ticket.

There were still a lot of hurdles to leap before deploying, but the major one was behind me. I was going. Everything remaining would be routine training and certifications. I couldn't believe it. The fight to be medically cleared was long and frustrating. Putting it to words was more difficult than expected. Deep down I had known I was still capable of doing the job but being told at every turn that I wasn't had taken its toll. Being denied three times had left me feeling so near to defeat that I had actually packed my bags just in case I was declined and booted from White Cell. All that energy spent getting letters and going through physical evaluations, all those hours, had failed to sway the med board. One chance to talk to a single person bluntly and honestly got the job done in ten minutes. Pretty sure if I had lied or sugarcoated my previous injuries, it would have gone differently. I just boiled over with frustration and blurted out how I felt. Aside from a minor lapse in military customs and courtesies with a lieutenant colonel, it worked great.

My attitude at MOB didn't change completely but my lower stress level and kinder demeanor were immediately noticed. With the knowledge that the squad really was going to stay mine, I called the boys together and explained what had been going on. The non-deployable status, the battle to be rid of it, the White Cell sneak, all of it. I wanted them to know me, sure, but more importantly I needed them to understand their "boss" gave a shit and wouldn't bail out given the chance. I've served under selfish leadership in the past and it is beyond demoralizing. Luckily I have also served under guys like Chewy and Bull. I wasn't buddy-buddy by any means. I demanded a lot and didn't offer up much for squad debate. I also frequently bit the head off any chump who decided to say anything sideways to my boys that they didn't earn. Strangely enough, with my focus no longer centered on my own deployment fears and issues, I really learned to love my guys. Feeling ownership for the well-being of my soldiers wasn't limited to my squad. Our entire platoon worked together every day with different squad members filling into different trucks. Everybody eventually worked with everybody else, and we were overpopulated with that special brand of weird. A guy shaved his eyebrows off entirely by accident. Two others hit either a base official's parked sedan or a tree with house-sized route-clearance trucks. For shits and giggles, another shaved exactly one hemisphere of his body hair off. One frequently donned a gorilla costume and terrorized leaders' briefings. Several more wore animal thongs and luchador masks when there was too much attention on the owner of the gorilla suit. The list goes on and on of the bizarre and child-like antics and eventually, the little sons of bitches got me. Baerga, Jones, and myself had just made staff sergeant, so I knew my turn to get the crap kicked out of me by a zoo was coming.

Most, if not all, units have customs they observe when a soldier is promoted. Back when rank was a little chunk of metal you pinned on, the brass backs were taken off and it was pounded into your skin through your uniform. The higher the rank you achieved, the more mangled your collarbone got in two small spots. When the army switched to fabric Velcro on rank, center-chest-punching became a thing. That caused physical medical issues and was frowned upon. Lasting bodily harm wasn't the point so when someone got a year older or a rank higher in our company, we pink-bellied them until they bled. Everybody loved being a slapper, but many cried foul when it was their turn as "slappee." Not this troll. I had been lured to the common room under the pretense that Staff Sergeant Jones wanted to go over some tactics for an upcoming training lane. No one said a word when I stopped in the doorway. A dozen guys in pounce position lining the room with several of them wearing animals on their genitals said it all. I charged with reckless and hopeless abandon, using my rifle to hold the gorilla at bay. Jones came from behind and put me in a choke hold.

My feet were no longer on the ground, but I could see eye to eye with an elephant. The next thirty seconds were a painful and hilarious blur resulting in a broken M4 optic, torn shirt, split pants, and a new fingerprint database on my stomach. While recovering my dignity, congratulations on my promotion were said in various ways, but the message I heard loudest was a simple, "Glad you're coming with us." I had known deploying meant a lot to me. Until my final doctor meeting and the promotional beating a week later, I hadn't realized going with my guys meant everything. Another month of MOB and we were off to Afghanistan. There were a few weeks of travel, additional training, and the process of moving onto base and taking control of all the trucks and equipment. All in all, we were living on Forward Operating Base (FOB) Fenty in Jalalabad, Afghanistan by December 2, 2010.

FOB Fenty wasn't a big place, but it did have an airstrip and most creature comforts. The base sat squarely inside the city of Jalalabad, in the province of Nangarhar. The airstrip itself divided the entire base into two halves. We called them Clean Side and Dirty Side, respectively. Clean Side had the chow hall, shoppette (like a corner store), airport tower, our motor pool and company Tactical Operations Command, base personnel living quarters, phone cafe, internet cafe—the works. Dirty Side had all the housing and shower points for the combat troops. To get to Clean Side, we either had to walk around the entire base on the only road aptly named Perimeter or walk across the flight line. Everyone just walked across. Base operations even installed a giant, red-and-green blinking light so soldiers knew when a plane or drone was coming in. Some of those drones were awfully quiet on approach at night. Every once in a while, you'd be halfway across the flight line when the green glow shifted to red, and an incoming Predator drone would nearly take your head off. The company Tactical Operations Command was located almost a mile from our hooches on the opposite side of base. I immediately foresaw a lot of walking back and forth in my future.

To house everyone, my platoon was assigned five Barracks Huts (B Huts) among a sea of them on Dirty Side. Our five huts were arranged in two columns: the first column had three B Huts, the second column on the right had two. Moving down column one and then down column two, squad assignments were as follows: Second Squad, First Squad, Third Squad, Fourth Squad, and one overflow hut because a single B Hut couldn't house an entire squad. Our five were surrounded by more B Huts but they weren't assigned to us so I can't remember who else was where, or how many there were in total, but it was a lot. Similar to shipping container storage yards, all of the B Huts were spaced about twenty feet apart end to end and ten feet side to side. A B Hut is generally an eighteen-foot by thirty-six-foot rectangular box made of plywood with corrugated metal for roofing. Both narrower ends of the B Hut had a door with a small air conditioning unit mounted above it; there were no windows. Inside, plywood and two-by-fours divided the space into three main sections. The middle section was a three-foot-wide hallway stretching from door to door. The left and right sections of the hut each contained four small rooms, again constructed with plywood and two-by-fours. With a total living space of about fifty square feet per soldier. Each mini room was lavishly furnished with a narrow, raised plywood bed frame complete with a used twin mattress. Almost all of the rooms had working power outlets. If an outlet didn't work, we tossed a cord over our makeshift walls since they didn't reach the ceiling.

Eventually, after enough tall folk disrupted my sanctuary, I nailed a bedsheet into the ceiling as a makeshift wrap-around curtain. Then I secured a rolled, razor-wire obstacle around the top of my divider walls and hung action figures by the neck with parachute cord to serve as a warning against future attempts to invade my cave. Each guy did a fair amount of customizing their space to hold all of their personal gear and make it their own. Build a shelf here, hang a lamp there, hammer in a few nails to hang body armor, draw on the walls, that sort of thing. I took the spot closest to the most-used door so people wouldn't have to disturb my whole squad just to get to me. I knew making myself easier to access was a horrible idea but liking and needing have never been the same thing. The hut I chose already had enough shelving to hold all of my gear so long as I combined socks with grenades. I was settling in just fine.

As a twenty-four-hour-a-day live-in boss, a squad leader can and does dictate just about every aspect of their soldiers' existence on a deployment. Changes to the daily schedule were more regular than the schedule itself, so squad leaders had to know where their soldiers were at all times. For my squad, we had a dry erase board with four columns. Everybody's name was listed in the first column. The second column was the "on mission" column and only had an "X" signifying that a guy was out on mission, or a blank space. The third column was a "write-in" location. If you weren't on mission, this column told me where you were. Things like gym, chow, motor pool, or poker night were written most commonly.

One particular maintenance day, the base had a blood drive. Just like any hospital, military aid stations and trauma centers have blood storage, just on a smaller scale. From time to time, soldiers are needed to refill those stores and you just get used to occasionally donating some of your precious red goo. My entire squad went to donate theirs around lunchtime and wrote "blood/lunch" down the third column, which is exactly what they hollered and sang as they pranced off across the flight line to chow. While they were gone, big leadership had come by for an inspection. Our commander was with a lieutenant colonel from somewhere. I forget the guy's name, but he was a nice enough fella. Smiled as he talked and walked among the barracks just checking things out, seeing how we lived. That demeanor rapidly dissipated when he got to my hooch.

"What the fuck is a bloodlunch?" was all that came out his mouth three steps in.

Now I knew it was written after each guy's name. What was new to me was where the dry erase board used to say, "Third Squad Whereabouts," now boldly proclaimed, "BLOODLUNCH." I put my entire hand over my face and took a long sigh. Smiling through my fingers, I replied, "My kids, sir." It was all I could mumble.

He'd obviously figured out what it meant by just being on the same base or looking at the rest of the board, because he chuckle-grunted in response and left without looking through anything. Best inspection ever.

The fourth column of the board was my notes section for each soldier—if they had mail, an upcoming shift of extra duty (like tower guard), mission roster changes, etc. It was easily the largest section of the board. I filled its entirety with, "I hate/love you assholes. This is our name now," and an arrow pointing up to someone's well-timed handiwork. I still don't know who wrote it, but it couldn't have been a better fit for a band of miscreants such as us. For better or worse, this was my family for the next while.

Chapter 5

Sometimes Bad Luck Is Your Only Enemy

We had been in Afghanistan a few weeks when Operation Black Panther came up. Operation Black Panther was a multi-day, named operation and it was chock full of firsts for the platoon. Our first IED find, first time a crane picked up my truck, first dismount to help civilians, and first IED strike on my truck were all going to take place. We were to go into relatively unknown territory. Reports were fairly thin, and it was unknown whether much of the route was even passable. Normally, when a route clearance mission was needed, we were notified of where and when to clear a specific set of streets and sent on our merry way, no support given, requested, or needed. The mission did not get a name. Named operations are another type of event altogether with multiple units conducting multiple missions at the same time toward one unified end goal. Infantry, artillery, route clearance packages, air reconnaissance, fixed wing (planes) and rotary (helicopters) air support—named operations get all the goodies. You only get goodies ahead of time if Big Army thought you were going to need them. The overall goal was to get to a small outpost in the Sherzad District Center and establish safe routes between there and FOB Connolly. As a whole, the platoon was nervous and excited as we moved from Fenty to Connolly to conduct rehearsals with the other companies and platoons we would be working with.

Our first IED find of the deployment, without help from the old company, was en route to FOB Connolly, about forty-five kilometers southwest of Fenty in the Khogyani District. Three quarters of the way to Connolly, an Afghan National Police truck came flying from off the road in a field to our right, drove directly into our lane of travel, and slammed on its brakes, quite effectively pulling us over. They had found an IED and had the trigger man wrapped up in their pickup. That meant the guy who was going to detonate the IED had already been apprehended and beaten to a pulp by the Afghan police. Apparently they disliked folks putting bombs in their roads as much as we did. Who knew?

It was Christmas morning, and we couldn't think of a better present than an IED *and* the guy who put it there! It was just too bad that our first find was "free chicken," or handed to us with zero legwork of our own. We dismounted and chatted up locals as we enforced a safe perimeter around the bomb, so still a win. Further in the plus column, the platoon was finally able to put all that training to use in the real world and pull a live bomb out of the ground without a single person getting hurt. It was midday when we finally arrived, and everyone was gathered in a large open expanse near Connolly's burn pit. It was the only space big enough to conduct practice runs with everyone present. We were the last group coming from another FOB, so once we arrived the brief began. Once the briefing was over, all soldiers involved in the operation took up their respective positions in our giant sand table and ran drill after drill, adding different ambush scenarios each time we finished. To run drills for our route clearance package, each truck crew stood grouped up in little gangs of five men in the order of march. We rehearsed for a few hours before everyone moved to their equipment and readied up to roll out. Mission start was supposed to be 1900 but because another unit was having trouble with a vehicle or something, we finally rolled out the gate around 2200. Right off the bat things looked to be only going downhill. Basically, the mission was to check out the area of operation for a few days and plan an eventual expanded military presence in the area. Our route clearance package actually ran into zero enemy presence throughout the whole mission and every group of civilians was very pleasant to deal with. A few other elements of the operation had some small arms engagements, but nothing serious and no soldiers or civilians were injured. Our portion of the operation was to find viable routes to and from strategic positions. That is a fancy way of saying we were used like guinea pigs to see if military vehicles could fit on the roads Big Army wanted to use as main highways for military traffic. Even though we encountered no direct combat, our piece of the pie was riddled with problems. Route clearance was for finding and neutralizing bombs, not exploratory driving. Our route, plotted by FOB personnel, took us down goat paths and mountainside trails. The terrain was unforgiving and slammed the truck around similar to throwing a computer down a flight of concrete stairs. At about the halfway point through a particularly rough pass, I looked down to find my feet swimming in a thick puddle of antifreeze pouring through the front firewall. It had obviously been going for a while because that puddle was sloshing around the entire truck, front to back. We pulled over near a canyon edge and hunted for the source. With the help of Walz, our mechanic in the wrecker, we found a busted hose, replaced it, and were back in fighting shape in no time. Though our mechanics had tools to fix major damage, they didn't have detailing equipment like wet/dry vacuums. Our now thickening and stinky sludge puddle was going to be with us for the remainder of the operation. Once on the road again, Sutherland noticed that a very small amount of fluid kept trickling in. Not enough to be an immediate concern, but it would need to be fixed on the next maintenance day.

A few hours later, while passing through another small but dense mountainside village, the weight of my truck crushed the road beneath us. We were going over a small land bridge with a mud wall on the left and a six-foot drop into a farm field on the right. The land bridge was maybe one hundred feet long, but about halfway through, the front driver side suddenly sunk about a foot. Sutherland brought the truck to an immediate halt.

Alarmed, I asked him, "What the hell was that!?"

Without looking at me he replied, "Uh, nuthin'?"

"Well, keep going then."

We made another foot of travel before the truck went from slightly leaning to lying at a forty-five degree angle about to flip over.

"That ain't fuckin' nuthin', Robbie!"

Now Sutherland looked right at me with a wide-eyed, *Oh, shit* grin. "Nope. No, now we are definitely in trouble."

It wasn't a life-threatening situation, but if we fell into someone's farm we would have destroyed it in the process of getting back out, and relations with the local populace were extremely important. At that point there was nothing we could do except get on the radio and let Lieutenant Campbell know we were going to need the wrecker to come up and recover us. Remember how I said land bridge? With no way around, and their position being far back in the route clearance package, there was no way to get the wrecker up to us quickly. We had to have all the trucks between us back out of the village one by one until the wrecker was positioned directly behind my truck. While the vehicle shuffle was being conducted, my crew slowly and carefully egressed the teeter-totter truck. We used the driver side door because mine was too close to the ground to open and the troop door was on such a steep angle that if we opened it, the truck would most likely flip.

I stayed in Sutherland's seat to use the steering wheel during recovery while the rest provided security on the ground. I'll be honest —they totally weren't performing any security duties. Lissy, Ballantyne, Heiner, and Sutherland mostly spent the evening talking with locals and making fun of the one guy idiotically still inside the truck. Seriously, I've had a lot of experience with the old "point and laugh" in my time and this was a whopper. Once the truck was recovered, we actually had to turn tail and bypass the entire village. During recovery, locals showed my boys three or four more spots that were tighter than the one we had gotten stuck in. Thanks, Big Army. Great recon job. Our plan for overnight rest was just to stay mounted and sleep in the trucks. It wouldn't be comfortable but sleeping outside was not a safe option. Around 0300 we circled the wagons and bunked down in a dry riverbed for a few hours of planned rest. Day One down.

Having accomplished almost nothing on Day One, Day Two's start time was pushed up and we rolled just before sunrise around 0630. As the sun began cresting over the horizon, a brilliant orange-and-purple sky greeted us. Then our hood inexplicably and abruptly erupted in fire. Each vehicle in our route clearance package had a tow bar, cable, rope, or strap mounted to it. Problematic guns and bombs that occasionally pestered us aside, trucks broke down all the time just from regular use and abuse. My rig had a big, fat, four-inch tow rope that sat just off the front bumper between the grill and the RPG cage. Unbeknownst to us, when we used the tow rope during our recovery the day before, we damaged some exposed electrical cable near its storage location. That frayed cable somehow came into contact with the tow rope while we were driving in the rocky terrain, causing it to ignite. Sutherland was in a decent mood.

"You know, for being a crap hole that hates us and doesn't always have running water, Afghanistan has some pretty impressive sunrises. You wouldn't thi—"

Whoosh!

"Oh, holy shit!"

In the span of two or three nanoseconds our view of the sunrise was eclipsed as our entire windshield turned into a blue-and-white flash immediately followed by red-and-yellow flame. Lissy's calm, yet agitated response was the best.

"Sun isn't even all the way up yet and we're on fire. Kind of day it's going to be, huh? Awesome. Let's put our fucking truck out and get on with it."

Every time the lead truck stops for any reason, be it to inspect a suspicious area, engage an enemy, or put out a hood fire, we have to call it in to the rest of the platoon so they know what's up. I was not looking forward to radioing this in.

"Six, Three, over."

The radio squawked Lieutenant Campbell's reply. "Three, this is Six. Go ahead, over."

I didn't want to go over the net and admit to being on fire because of my own stupidity, but what else could I have said? A fairy shit on my hood and her sparkle poo engulfed my hood in magic flames?

"Roger, Six, we're having technical difficulty at the moment and we need to halt for just a few, over."

Usually, I supplied more info about a reason for stopping, but I gambled he wouldn't ask for further information if it wasn't a long stop. It wasn't short enough.

"Three, Six, what difficulty are you having, over?"

Damn it.

"Roger, Six, our hood is experiencing a Fourth of July electrical fire. We need a minute."

Lieutenant Campbell was not pleased. "Three, say again. You are on fire right now? Over?"

Sutherland and Lissy were now standing on the hood of the truck beating the flames with a fire blanket.

"No, Six, it was just a small spark or two, nearly handled."

The reply came through in a brusque voice. "Three, I'm gonna need a full situation report when we get to the next rest location."

Damn it again! Spoken quickly enough, my reply sounded like an embarrassed high-pitched one-word mumble

"Roger-Six-no-problem-Three-out!"

Shortly after that exchange, the fire was out. Since it had burnt nearly to carbon, Lissy tossed the tow rope remains into a deep ditch to hide our shame and hopped back in the rig.

"Six, Three, we're good to Charlie Mic."

Charlie Mic was code for continue mission. Away we went with a three-foot-diameter scorch mark where our tow rope used to be.

The rest of Day Two was pretty much smooth sailing. There were only a few kilometers left to clear before reaching the next rest site and we desperately needed a real break. The sludge and its now noticeable aroma didn't help. Occasionally, we got out of our truck and stretched our legs, but that was for a few minutes and only when up-close investigation was needed or a blocked path was to be cleared. When the need to relieve ourselves arose, instead of going to a rest stop or gas station, we peed in bottles. Yup, we'd simply drink a Gatorade or bottle of water then fill it right back up. As I hate unnecessary litter, I never allowed my guys to throw their bottles outside.

By night two of Operation Black Panther, we had amassed a small assortment of half-full bottles. It was also December so we had our heater going full blast. An RG truck does not have air conditioning or heating in the back, so to keep Ballantyne, Heiner, and Lissy from turning into popsicles, we ran black plastic tubing along the sides of the floor as makeshift vents. It worked out great and kept our truck nice and toasty.

Around 2330 everyone in my truck heard a low boom and thought it was a hit, an IED strike. The mood instantly shifted from tired and weary to frantic and alert. Obviously it wasn't us, as nothing inside the vehicle had quickly or violently changed its position. I barked orders.

"Lissy, Ballantyne, look out back and tell me who was hit! Heiner, get on thermals and find me a trigger man. Robbie, turn us at an angle and block the road to keep civilian traffic from getting in."

I reached for the mic button on my headset to get on the horn to the platoon but was hit by an impassable wall of hellacious stench.

"What the—*huhgohlk*—" I began choking on air and every word was labored. "What-is-that-SMELL?"

No one was capable of answering my inquiry. Our eyes watered; our coughing and dry heaving was uncontrollable. Sutherland damn near kicked through the firewall again when he hit the brakes to keep us from careening off the road. After a few seconds of being unable to speak, Heiner finally piped up over the VIC3 radio system and confessed.

"So... one of my pee bottles somehow—*muhguhgamulk*—managed to wedge itself into one of the, uh—*gugh, gugh*—heating tubes. I guess it was hot enough to—*huhblurpff*—burst the bottle, Sergeant..."

When it blew, it sprayed ultra-hot and stinky urine all over the floor of the truck and vaporized quite a bit into the air. This, combined with the gunk already on the floor, made the smell almost unbearable. We cracked the roof hatches, shoved napkins, socks, or bullet casings up our noses, and continued on our adventure. There was a slew of yelling and verbal bashing of Heiner over the next few hours, but Sutherland summed it up perfectly. Tears rolled down his frown-covered face, his hands were white-knuckle-grafted to the steering wheel, his head and eyes fixed straight forward—he never even turned around.

"Well, this is fairly awful. Only a full twenty-four hours left cooped up in here. Joyous day. I fucking love *and* hate you, Heiner."

If you aren't aware, similar to tear gas, prolonged exposure to the aroma of burnt pee will sear your nostrils and make you cry. It's the very definition of yucky. When we arrived at the rest site around 0200, every door and hatch flew wide open. Even in the chill of a December night, the doors remained open until we rolled the next morning. Day Two down.

The final day of Operation Black Panther was short and sweet. Okay, it wasn't short, but we had zero issues on the road, and only had one extended stop to help out an old civilian fella who managed to get his faded white pickup buried in some mud. Our truck still smelled like hell, so we all took the opportunity to get out, except Heiner. He remained inside, manning the gun with his stink. It took longer than it should have because none of us wanted to get back into the damned pee rig, but after about fifteen minutes of playing in the mud while pretending to scratch our heads about the predicament, we had to get back on the road. Plopping around in the mud was the highlight of Operation Black Panther for me. So with a little sorrow in our eyes, we hooked the kind man's little mud boat up to our giant monster using the Husky's tow rope. Tugged him right out in seconds. With no more excuses or incidents, we mounted back up and actually made our way to FOB Connolly at a reasonable hour.

The original plan was to stay at Connolly overnight, rest up, and go back to Fenty in the morning. However, we were all sick of being in the trucks and with no mission scheduled for the next day, the platoon opted to grab to-go boxes from the chow hall and just roll on home to Fenty. By the time we got into Jalalabad, it was past 2200 and everyone was out of go-go juice. I don't know about the other trucks, but my crew was struggling to stay awake. Hell, the guys in the back had been passed out for a while. As we neared the base I, too, drifted in and out of consciousness, calmly being rocked to sleep by the clunk and thud of our poorly operating truck. Until one last thing decided to go belly-up on good old Operation Black Panther.

Less than one kilometer from the front gate, Sutherland joined our siesta fiesta and crashed us right through a power pole and the concrete median it was mounted on. Being the last one to stay awake, I don't blame him at all. We had been going hard with little to no respite for several days. Our rig effortlessly climbed the oversized median and plowed through the slightly narrower concrete base of the pole as if it were a stunt ramp. A transformer on the left side of the street simultaneously blew up and along with it, all the lights on the block immediately went out in a blast of sparkling glory. For a split second, I bet our airborne vehicle looked pretty majestic in the night sky surrounded by hundreds of little concrete meteors.

All of us were jolted back to a delirious state of alertness as the truck's front tires came crashing back to earth. When our fender-bender occurred, we had been in the process of making the final turn before reaching Fenty's Entry Control Point. No other truck had eyes on us during the crash. All any of them saw and heard was a huge burst of light a block away and a loud explosion. I again hopped on the radio and explained to the platoon that it was not an IED and that we were physically okay, but that we would once again need the wrecker's assistance. We had demolished the entire front end and our steering function was gone.

Amazingly, we could still drive forward and were facing almost the correct direction. The last few hundred meters to the gate was a limping and leaking struggle of epic proportion. Upside: Since the engine bay was now leaking different fluids from basically everywhere, the source of the mysterious sludge puddle was no longer a concern. With pieces of vehicle literally falling off all the way to the gate, we had made it back home. From there the wrecker pulled around front, hooked up, and towed us through the serpentine. Our first big operation of the deployment was over and my truck, which was now demolished, had caused more problems for US forces than our supposed enemies. It was a terrible start, but it was a hell of a learning experience.

Operation Black Panther honestly wasn't very unique from other missions, except it was three days long instead of ten or twelve hours. Common problems were rollers, tires, and axles being destroyed trying to get through impassable terrain; roadways breaking, resulting in vehicle flip-overs; normal break downs; and bullets, rockets, and bombs of varying sorts. The Buffalo, our biggest rig, once tipped over into a ditch simply because of slick mud. That early morning mission ended well past 2200 because recovery took so long. Eventually we all became really good at just rolling with whatever pee-scented sludge puddle came our way. Which was good because as months progressed, spring brought much more trouble.

Chapter 6

Spirits Only Stay Broken by Choice

Weeks turned into months, and everyone settled into a rhythm in their respective roles. Especially Sergeant Byron Lissy, who was originally from another unit and had transferred in. I first crossed paths with Lissy in 2006 and we did not immediately get along. We met when I joined the unit in Kuwait and got right to huffing and puffing at each other. He stands about six feet tall and is stoutly built. He wore his thick, curly black hair about half an inch long in a "within regulations" mohawk and an ultra-thin matching mustache, all expertly coiffed. He's got dark skin and a band of freckles that stretch underneath his stern, brown eyes. It's difficult to pull off both the pretty boy professional *and* ferocious as fuck soldier looks at the same time, but he managed it. From the neck down he was pretty much a furball. I even threw a little Velcro ball at his chest once and it stuck. He was hairy.

We had served in the same squad in Ramadi, but by the end of that deployment he had been promoted to my squad leader. He was my third one that tour. Chewy (Sergeant Schoenthaler) and Bull were still alive but had been sent home from injuries. Neither one of us can pin a specific reason, but my guess is we clashed at first because he had every hair in place, clean, crisp uniforms, and really looked the part of soldier. Me? I resembled a living pile of compost squished into a uniform. I was replacing a soldier who had been injured during the last two weeks of training. To be able to join the unit in time, I received an expedited one-week mobilization. The hours were long, and sleep was not really much of a thing.

Slowly our inexplicable anger posturing turned from competition and mutual disdain to collaboration and admiration. As people were injured and we went through rougher and rougher missions together, it changed again from collaboration to working side by side with a new brother. We operated in the same truck often. Occasionally it exploded. Those memories are important, but one of my favorite memories of Lissy were food challenges.

Again, pesky boredom would surface, and we would challenge each other to eat random, gross things we found. The only rule was you couldn't challenge someone to do something you yourself would refuse to do. This kept ridiculous challenges like "eat that rock" off the table. Once, he tasked me with drinking an entire bag of old cooking caramel. If you don't know what that is, cooking caramel is not like the individually wrapped little bits of heaven you get at a candy shop. This stuff is an oily, chunk-filled, gooey mass. Essentially it was a caramel/oil lava lamp in a gallon-sized bag. I drank it in under five minutes and threw it back up in even less time. I was in luck though, and my revenge was truly evil.

Halloween had come and gone a few months back and I found a two-pound bag of candy corn shaped like pumpkins in my footlocker. A time limit was always involved. With the vomit fest that had attacked me after the caramel, I knew a few minutes was too short. Don't remember what film it was, but he had from the start of the movie until the end to eat the entire bag, ninety minutes, give or take. He confidently set up his folding chair in my room and began eating. It was an action film and just after the first mind-bendingly, well-choreographed fight scene, I peeked over to see Lissy's progress. He was no longer his normal shade of healthy brown but had gone a kind of greenish-gray. Bits of chewed up orange-and-yellow candy wax dotted his extremely unhappy face.

"You gonna make it, man?" I chortled with a mild amount of concern for my frog-colored friend.

As if he had been drinking for hours he responded with a slur. "Doooin' gooo... I'llph beee f'nine"

He was not fine. We finished the film, and he heroically finished the bag. He then went out back and redecorated a good chunk of the base in candy wax vomit. After what felt like an eon, Lissy stumbled back in, looked at me with one operational eye and chuckled. "Expired candy. Not bad, Leprechaun. You'll get yours, but not bad."

Along with rough missions, pranks escalated pretty steadily after that. At the height of our mischief we began ambushing each other with BB guns at the most inopportune moments, such as sleeping or shower time. The interior walls of our "rooms" in Ramadi were sheets of plywood that did not reach the ceiling. Lissy's raids were legitimate cause for sleeping with eye protection on and a groin plate covering your nuts. If you were to ask him today, Lissy would emphatically declare that sleep raids did not occur. Like a rug, he lies.

Even though Lissy lived seven hundred miles from me we kept in touch after our time in Ramadi ended. In early 2010, my unit needed additional bodies to fill our roster for the upcoming deployment. Leadership asked some of us if we knew any capable soldiers from the Ramadi tour who'd be willing to transfer. Lissy was number one on my very short hunting list. Though he wasn't in the 744 proper, Lissy was in the same battalion, so the logistics of the transfer were not an issue. To get him into our company and be put on the roster, all we had to do was get some paperwork signed by both company leaders. The tough part was to be able to convince his family to let him go again. He was engaged to his future wife, and they were making pretty big life plans of their own. Deciding to volunteer for another deployment is not the same as the army telling you to go, especially not in the eyes of our civilian families. After a few days of both he and I making assurances, explaining why it had to be him, and some groveling, he got the go-ahead. He took a couple days to get his affairs in order and say his goodbyes, then joined us.

My squad couldn't have been luckier. Lissy is a physically strong, intelligent, and quick-thinking man with a lot of experience to give. Don't recall what his title on paper was but he was the Mama Bear of our squad. I was still new to leadership and my edges were not smooth at all. His aid or instruction always came from a peer's perspective, and he treated my guys respectfully until, in a very short time, they were his guys too. I always listened to my soldiers when they needed an ear, but I wasn't much kinder than say, Oscar the Grouch. Learning to be kind while leading is what Lissy taught me. Not the bake-you-some-cookies type of kind but understanding your soldier's fears and concerns from their perspective and showing them you understand because you're suffering right there with them. In the months training at the mobilization station and on mission in Afghanistan, Lissy proved over and over to be an invaluable brother. Not to me; our bond was already established. My kids, however, weren't in Ramadi. Couldn't say exactly what it was that he did as I was going ninety miles an hour in every direction, but it was like I looked up one day and Lissy was accepted as family. It felt as though he had somehow always been that way with the whole company. Guys from other squads asked him for help, and he attended leaders' briefs and brought valuable insight. Lissy had become the vulnerable point search lead. He and a small crew cleared surprise hazards, spoke with locals to clear traffic, and searched IED threat zones that machines couldn't.

Then came BOW.

BOW stood for Bastogne (and, later in the deployment, Bronco) Over Watch. It was a monthly resupply mission along Route California where fifty or so semi-trucks full of things like food, water, ammunition, and mail were delivered to a remote base called FOB Bostick in the northeast corner of the country. Our piece of the pie was to clear the road from Asadabad on up to Bostick ahead of the delivery trucks while they were escorted by an armed convoy from another company. This monthly operation was where we took a large portion of our casualties during the deployment. My platoon would pack up in FOB Fenty and head the ninety-two kilometers to FOB Wright near the city of Asadabad, where Second Platoon was headquartered.

The first night was always low-key. The guys from First Platoon hung with Second and caught up while all the leadership went to briefings about the upcoming op. To keep things from becoming too predictable, mission start times were never the same as the month before and we had staggered route clearance coverage. One platoon would roll out and begin clearing to Bostick, the other would roll a few hours later with the semi-truck convoy and its escort following shortly behind. Some months we rolled first; some months Second got the honor. Some months only one of the platoons took mild contact and other months were an all-out firefight shit show for everyone.

Route California is, for all its insanity, a beautiful stretch of road, winding alongside the Kunar River while rising higher and higher through Kunar Valley's Lower Hindu Kush Mountains. Once up into those mountains, views of enormous, stark-gray rock faces are interspersed with lush, green vegetation and the mighty Kunar River itself. Every few miles a multi-terraced farm is dug right out from the mountainside supported by a few small, one-story stone/clay buildings. Where the river is wide, these small farms turn into villages. Where the river is narrow, it looks as though humanity has never been there before.

Route California is paved for a good portion of the trip to FOB Bostick, beginning in well-populated valleys and towns. But the more than twenty kilometers of mountain pass itself is dirt. Most of the dirt section is wide and easily traversed provided there aren't any bombs. Every rainy season, however, erosion ate up little sections of the road and portions of the pass became extremely hairy. Once, on a BOW that had gone late into the night, Sutherland was struggling seeing and wanted me to visually confirm how much of our right front tire was off the road. This was discomforting because Robert Sutherland is one hell of a driver. His particular use of the words "off the road" wrenched my stomach. To our left was a sheer rock wall, which he was already hugging so tightly he broke our side mirror off. On our right was a sixty-foot plunge into that pretty river mentioned earlier. I squeamishly opened the six-hundred-pound door using the hydraulic assist button. As it lurched open, I leaned an uncomfortable percentage of my body over the edge to see past the chain ladder and got a good, long look at the tire.

To my dismay, all I could holler was, "Seventy percent still on! No more right, NO MORE!" I desperately tried to keep from shitting myself while staring into a dark blue abyss that was too far away to hear over the drone of our engine but close enough to see water as it crashed over rocks, making big, white foam. Whether from bad driving or ambush, in multiple locations along the route, there are burned-out vehicles rusting on the side of the road or in the rushing waters below.

Occasionally along California, there are overwatch positions. These are just one or two trucks with guns mounted, parked, full of soldiers in historically hazardous areas. As you get higher into the pass there were fewer and fewer simply because only one vehicle would fit on the entire road section.

It was another breezeless, sunny afternoon on March 21st, 2011, and around one p.m. we were nearing the 883 grid line, or the Shark's Tooth, as we called it. This was a particularly beautiful section of the valley that semi-flattened for a few kilometers. The river was wide and there were green farm fields with huts dotting the entire mountainside. Route California is on the left side of the Kunar River. Across to the right side was a sharply pointed and steep hill covered in vegetation too thick to walk through. It was several hundred meters tall and looked like a giant, green-dyed, great white tooth.

Even during the more peaceful BOWs, this section of the run was always eventful. Small arms pot shots and the occasional RPG were pretty regular. Two guys from the company we replaced died right in the same area from a complex attack only six months earlier. Dismounts were already on the ground performing sweeps off the side of the road and ahead while the vehicle team cleared the road surface itself. I can't recall what direction the shots came from first, but they came from three separate locations at almost the same instant. Two-thirds of the fire was directed at dismounts; the rest peppered the truck team.

Bad guy Team One was shooting at us from across the river with a slew of AKs and one ZPU2 (double-barrel, Soviet-made 14.5-millimeter heavy machine gun). Bad guy Team Two was north of our position on the outskirts of a very small village. I don't know the name of it because it's too small to even be recognized on a map, but it was pronounced Juh-lay-luh. This team of bad guys were using all smaller arms and some RPGs. Bad guy Team Three was actually on the mountainside a few dozen meters in front of our dismounts.

My truck was lead so we suppressed the hostiles near the village while the rest of the trucks engaged the ZPU2 team across the river. Dismounts were much farther off road and were basically on top of the third enemy group. If I recall, it was only three or so enemy personnel and the dismounts were able to permanently quiet them very quickly. The ZPU2 team proved to be the biggest threat as they also had some 7.62 Dragunov long-range rifles. About the same time that bad guy Team Two in Juhlayluh was handled, the river team increased its rate of fire exponentially and began laying accurate shots.

Lissy was hit while trying to come down the mountain and regroup at the trucks. The bullet entered his thigh, blowing a large chunk of flesh off, abruptly turning up after hitting bone, and finally lodging in his abdomen near his spinal column. The radios were absolute chaos. I mean, Chaos was our company call sign, but this was surreal. Dismounts were all adrenaline, screaming out enemy positions and informing us that the enemies were regrouping, truck teams were coordinating fire barrages, and platoon leadership was on another network calling in fixed-wing support to come drop big-ass bombs on these enemy fortifications. After about a half hour of intense fighting across the whole platoon, the radio calmed down enough for Lissy to break in over the traffic and let the platoon know he was injured and needed help.

Calm as can be, Lissy's voice came over the net: "One Three November here. I've been hit and cannot walk. Gonna need some help getting outta here."

It's strange because from that moment forward, the net chatter became very clear and precise. We had an injured brother and all of a sudden winning the fight was second priority. Orton came up on the radio next and all he said was, "Okay, on our way to you now."

Getting to him was not as easy as saying it though. In the end, I believe it was Brummett and Pratt who got him down the mountain. Orton had to change course mid-ascent and go get Ammon, who had also taken a bullet in the thigh moments later. Ammon was lucky and had no major bullet travel, keeping his damage localized. Lissy, on the other hand, had gone a greenish-gray color and lost a lot of blood. Constantly awake but in and out of it mentally. He and Ammon were loaded into the CASEVAC vehicle for trauma aid while the MEDEVAC bird was called in. During the casualty-loading process, Hammon provided cover fire and received a through-and-through gunshot wound to his left arm. Leavitt, the CASEVAC driver, was out of the truck and helping load the men when an RPG detonated on a nearby rock and littered one side of his face, neck, and body with shrapnel. Leavitt helped Hammon into the back then hopped back in the driver seat.

When his truck commander, Staff Sergeant Jones, saw his face and made a slight expression of horror, Leavitt simply asked him, "Is my face still there?"

With a kind of *Oh boy* shrug, Jones just replied with, "Yup, looks great. You're good."

As the last casualties were loaded up, all the trucks buttoned up and the fixed-wing support we called in earlier dropped a 500-lb. bomb on top of the remaining ZPU2 team across the river.

With enemies down and our injured recovered, we made our way to an open field just off the road and loaded our boys onto the MEDEVAC chopper. When the flight medic saw Leavitt's face, he made a stink about him needing to stay on board. Since Leavitt was a bit combative about being MEDEVACed himself, we just pushed his ass onto the bird right before it flew away. We loaded back into our trucks and eventually reached Bostick, had a hot meal, and slept a few hours. Once all the semi-trucks arrived at Bostick we began clearing our way back down to Asadabad. Though it did occasionally happen, most combat was incurred on the way up to Bostick. The ride back down was quiet and uneventful.

Once we were all the way back home to Fenty, my additional assignment began. When we first arrived in-country, some extra duties were handed out. Extra duty is exactly what it sounds like, jobs that needed to be done and had no soldier already assigned to do it. If assigned an extra duty, all soldiers are still responsible for all their regular work duty. I had been given Summary Court Martial Officer. It sounds much fancier and more official than it actually was. Essentially, I inventoried and returned to their rightful place the gear and personal effects of injured and fallen soldiers. Bright and early the morning after getting back to Fenty, the commander and I went through Lissy's entire room. We had a box for personal items to be sent home to his fiancée, a box of gear to be turned over to base personnel, a box to be sent back to the home unit, and a box to burn. After several hours of cataloging and emotional turmoil, we loaded up the FOB runner, a topless, flatbed Humvee used only on base, and delivered each parcel to its proper place. It was already dark when I finally parked the FOB runner at HQ and began walking back to the hooches.

That's when it sunk in. Ammon, Hammon, and Leavitt's injuries were all small enough that, with a fair amount of arguing at doctors, all three men were allowed to stay and recover in-country. They would be on light duty and restricted to the base until a medical officer signed off on mission readiness. Lissy, however, was done. I was pretty sure he'd live, but he was flown all the way to Germany for stabilization surgery, so no one really knew for sure. A few days later we had a milk run of a mission and finished early. As we drove into the motor pool, the commander was waiting with good news. Lissy was up and walking and if we wanted to stick around and sit in, he was about to call the hospital and try to speak to him. Nearly the entire platoon bull-rushed the office and huddled around the phone. Captain Sparrow dialed the numbers. After a few minutes of transfers and "see if he's awake" checks, Lissy was on the line. He sounded upbeat but his voice boomed far weaker than normal. Everyone gave well wishes and condolences, but there was a tension in the air you could cut with a knife. It was nearly unbearable, so I thought to myself, *screw it*, and just asked the question that no one but me apparently was thinking.

"So you don't have to answer, but we're all wondering, you know... with one of your balls being shot off, are you all black or all white now?"

Definitely no one but me was pondering that particular question. Angry and aghast, *What the hell is wrong with you?!* looks were coming at me from every direction in the room. I said nothing but pointed to the phone and gleefully mouthed, "Just you wait," over and over. After all the guffawing settled, the faintest, wheezy laugh could be heard. When he, too, settled, Lissy's response vindicated me.

"Well, I got a pimp limp and a cane, so I'm saying all black, you little troll."

The mood lightened. I told my brother I loved him and ducked out of the room to think. Humor was my way of dealing with things and I tried to stay optimistic because Brown Bear was definitely going to live. But his career as a soldier was over and in the coming days that weighed heavily on me. I had convinced him to come on this tour, begged his family, promised I'd keep him safe, and instead he'd received a career-ending injury. Very quickly I shut down socially. If we weren't on mission, I was locked away in my hooch wallowing in guilt, barely eating, certainly not helping my kids get through their hard loss.

By about the fifth day of this, Orton (Fourth Squad leader) had had enough. He came into my barracks and banged on my plywood door until it nearly splintered to see if I wanted to go get chow. From behind what was left of my closed door, I thanked him and declined. However, he continued and after several more structurally endangering slams of his fist, I relented for fear the door could take no more.

Orton and I also first met in Ramadi. He stood just over six feet, had thin-frame glasses, and large-frame shoulders. Liked by just about everyone, Orton was a pleasant and courteous guy, except for his pipes. Not some kind of smoking pipes mind you, bagpipes. He played them whenever he could find the time, which, due to our job, varied greatly. We'd seen each other before, eaten at the same tables, exchanged hellos as we passed each other, but he was in Second Platoon and I was in First, so we didn't talk all that much.

The first conversation I remember having with him was late one morning after mission in winter 2006-2007. I had bunked down for some sleep and was just beginning to feel myself drift away when the shrill blare of Highland bagpipes shot me right up to fighting-condition ready. In just a pair of army-issued PT shorts, I ran out the back door of my platoon hooch to find Orton standing in the space between his platoon's living quarters and mine jubilantly squeezing a bag with tubes on it. I don't know the name of the song he was playing but I'm pretty positive it was the sound a banshee makes when they die.

I stuck my finger out at him and yelled, "I was almost asleep. Knock that shit off!"

He smiled as he looked down and retorted, "When am I supposed to practice, guy?"

With my finger still in his face I grimaced. "Good point. I'll put plugs in." I turned around and trod right back to bed and struggled to be serenaded to sleep by the Braveheart-ish soundtrack. But that's how we became friends.

Now here he was trying to get me out of my funk. He conducted the door-banging ritual multiple times a day every day after that. He did it for every meal, to go to the gym, go to briefings, to clean guns and watch a movie, anything at all really. As long as he was around, there would be no time for pity parties. Which was exactly twenty-five more days.

When we first arrived in-country there was an issue with having enough vehicles. Several of our company trucks had been burnt to the ground during a base attack a few days prior. Our next BOW was scheduled to kick off on April 15th, so a day early we rolled on up to Second Platoon's base and settled in for the night. Once we had the trucks off-loaded and housing for the night set up, we released the Joes. The walk from temp housing to the permanent side of the base was strangely peaceful with narrow, winding walkways. I say strangely because at night you had to walk using a dim, red light due to the base always operating in lights-out mode at night in response to "deter further accuracy in future night attacks."

FOB Wright was tightly tucked up into the mountainside a few miles outside of Asadabad. Driving through one of the entry gates felt similar to pulling into the parking lot of a combat-themed amusement park bolted to the side of a mountain. Tall stone and concrete walls adorned every side not looking at a huge rock face. Once through the gate, a winding road took us to the back of the base nearest the mountain where a spacious parking lot of dirt and gravel awaited us. There were concrete bunker/barriers and concertina wire rolled out for defense and to organize parking. I only ever saw twenty or so vehicles, but the area could easily accommodate ten times that. The back of the lot was protected by the mountain itself with a steep ridge rising hundreds of feet out of the ground as if to emulate a foreboding "no exit" warning sign. On the other side of the lot, once through the barriers, Wright was a pretty nice FOB. Stone buildings were stacked in on each other with narrow, winding walkways to get between them. Being on mountainous terrain, the base itself sloped heavily and had multiple little snow melt/rain ditches all along the walkways. I never really did figure out the layout of Wright, but it sure was pretty. Plus their chow hall was epic. Pancake-wrapped breakfast sausages were life-saving.

As soon as everyone had some of Wright's delicious grub in their guts, the leadership pieces headed to the mission brief. After becoming glaringly late getting lost along one of the very walkways I admired, I found the chapel and popped in to an already-in-progress briefing. Chatter was that something was coming the next day. We had Low Level Voice Intercept reports of insurgents mounting a considerable attack. I don't recall all the details, intel, or concerns that were discussed. I only remember bitching about the decision to use the same fallback MEDEVAC location as the previous month. We all thought it was stupid to land a big, defenseless helicopter in the same place twice, but that decision was above any pay grade in the room. EOD was embedding a truck team of their own consisting of Staff Sergeant Bailey, Sergeant Bigby, and their First Sergeant, whose name escapes me. Having an EOD truck team was fairly common for missions with a high likelihood of IED finds. We had worked with Bailey and Bigby before. After the briefing, I chatted up Bailey for a few minutes and then left to repeat a condensed squad-level brief with my boys. We were to roll early in the morning, so everyone bunked down fairly early. Only a handful of us stayed up the extra twenty minutes waiting for midnight chow.

The next morning was smooth and easy getting the trucks up. Usually with this many puzzle pieces in play, things simply go wrong. All I recall is a mild radio problem that our commo guy, Specialist Touzani, had fixed again with a half hour to spare. The boys brought to-go boxes full of my favorites while I plugged the route into the Blue

Force Tracker and double-checked that it matched my maps. In the early morning chill, the clouds were lower than the mountain peaks. We rolled out just as the sun was waking the rest of the valley. The day was calm, nearly windless, which was unusual given the fact we were in a mountainous river valley.

It took a little longer than normal to get out of town as Asadabad was bustling with people walking, people in cars, people pushing carts full of wares to sell—even cows were out and about. By noon we were completely out of populated areas and had switched to slower-moving, deliberate clearance. An hour past that and we were nearing the Shark's Tooth. This is where the intercept report had indicated a possible complex ambush would be waiting so we halted, dropped out dismounts, and began very slow and deliberate clearance.

Thirty minutes later we were rounding a bend when a muffled firecracker hit the windshield followed by a couple layers of ballistic window glass breaking. Robbie brought the truck to a quick halt while angrily sighing, "Someone is shooting my truck, Nelsoooon!"

Specialist Nelson, or Ogre, as I referred to him, was up in the turret manning our M2 "Ma Deuce".50-caliber machine gun and couldn't see the shooter. It was the middle of the day and the nearest cover the shooter could be hiding in was hundreds of meters away, so searching for muzzle flash was difficult.

Without a distance, direction, or number of shooters to report, all I chimed into the net with was, "All elements, this is One-Three. Be advised we're taking fire from what appears to be Juhlayluh and the Shark's Tooth."

Other trucks confirmed they were also taking fire, but the next few seconds of radio traffic was mostly gunners and anyone able to provide directional aid to them. While this was still going on, one of the dismounts broke in over the net.

"One-Four is down."

One-Four was Orton's call sign. Robbie immediately began moving the truck up while he and I did our best to visually clear the road as we finished rounding the bend and moved in line with the dismounts. I quickly broke in to let the dismounts know the plan.

"This is One-Three. Bring him straight down to the road, we're coming to you."

Dismounts only get a hundred, maybe two hundred, meters ahead of the lead truck so getting to their lead element didn't take very long —a minute and a half, tops. Getting Orton, unconscious and heavily laden with gear, down the mountain took longer. While Nelson and I coordinated spotting and shooting across the river, Robbie wanted to head up the mountain and help get Orton down to us. His civilian job was EMT, and I agreed with his logic, but thought it should be me.

"You could go, Staff Sergeant, but..."

He then pointed back and forth a few times to the two bullet holes in my window and the zero in his. I relented. Robbie pushed the hydraulic assist, hopped out, and headed toward Orton's position. He got about ten meters from the truck before a spray of machine gun fire landed directly in front of him, cutting him off. Robbie took cover behind a large rock and decided to fall back to the truck. While running back to us, a second hail of bullets landed on our truck, cutting Robbie off again. He retreated back to the safety of the large rock but was again cut off from going uphill. Every time he tried to go up the hill or toward the truck, he'd damn near get shot to pieces and it was visibly frustrating him. His back-and-forth bullet jaunt actually occurred several times in the short span of a few seconds. Amid all the chaos at hand, it was kind of hilarious. Robbie had had enough and during a lull in the shooting, he ran for the troop hatch at the back of our truck and hopped in to begin prepping to receive casualties.

While Robbie had been stealing all the bullet attention doing his back-and-forth death dance, Hartzell, the medic, and Pratt, the dismount sniper, had also made their way to Ballantyne and Orton. Ballantyne and Hartzell carried Orton while Pratt carried Orton's gear and covered their movements. The four of them were nearly to the road when Stoof, the driver of the vehicle behind mine, caught up to us and used his truck to provide cover loading Orton into the back. Pratt returned up the hill to rejoin the dismounts still fighting while Hartzell, Sutherland, and Ballantyne kept working on Orton.

I hopped in the driver's seat to turn us around and get to the MEDEVAC fallback site, which was behind us. Flipping around took a few hairy moments as the road was moderately narrow and any jostling of the truck made rendering medical aid difficult. As we got our nose faced the right direction, a loud but distant explosion went off. All of us inside heard it but couldn't tell where it was. We had turned around and were now facing the rest of the platoon, but aside from my truck and Stoof's, they were all still back behind the bend in the road. Nelson and I were the only ones with headsets still on so we relayed information as it arrived. EOD's truck had been hit by a large IED and flipped over. There were casualties and the truck carcass was now blocking our path to the MEDEVAC site. As a group, we decided adding another truck to the recovery effort was a bad idea and resolved to stay where we were until a path was cleared.

Enemy gunfire noticeably increased after EOD's hit and recovery took quite a while. I manned the radio and occasionally repositioned the truck for Nelson, who had been shooting the entire time. During a lull in the radio traffic, Nelson's deep, gravelly voice piped in to let me know he was nearing bingo on ammo.

"Hey Sergeant, I'm gonna need a plus-up soon. Almost black on ammo."

"How close to bingo are you? We gotta reposition the truck to get to it."

Ogre fired one more ten-round burst across the river immediately followed by his wimpish ammo count.

"Uh... I'm out."

"Mother fuc—" A maniacally frustrated smile/grimace formed upon my face and my control box was nearly crushed when I keyed the microphone to reply. "I'll go now! Use your fucking pea shooter until I'm back."

With loud, complaint-like noises emanating from my person, I ripped my headset plug out, angrily crawled back over to my side of the truck, opened the door, cleared off the brass and links that had piled up, and jumped out. From front to back, bolted directly onto the side of the truck's armored, V-shape hull were tall and deep storage containers full of food, water, spare truck parts, and ammunition. We had fifteen hundred rounds inside the truck and another thirty-five hundred in storage compartments on the outside. They were large and there were six on each side of the truck. Luckily, most of our .50-caliber ammo was stored in the closest two bins to my door. As the bottom lip of the door to the truck sits six feet off the ground and I am five foot six, throwing the ammo cans up onto my seat to get them into the truck was a pain in the ass.

Fifty-cal ammo cans are heavy and each one only holds a hundred linked rounds. After getting eighteen hundred-ish bullets onto my seat, I stood on the door sill, handed a few cans directly to Nelson, and stacked the rest at his feet so he could easily reach them. Bullets had been pinging off the truck and kicking up miniature dust clouds off the road the entire time this went on. I really couldn't hear much of anything over my loud bitching about Nelson and his impeccable damned timing. With that chore completed, I finally closed the door and was able to plug my helmet back in with a calm-ish demeanor.

Nelson apologized for not giving me any warning. I wasn't really mad at him as he was just doing his job—quite well, I might add.

"It's fine, Ogre. Make the bad guys be dead, we'll be even."

A few ammo cans later and word came over the radio that the way for us had been cleared enough to get to the MEDEVAC site and the bird was on its way. Robbie stayed in back while I slowly drove through the rest of the platoon to meet up with our medics. It was slow going because at every truck we had to halt and wait for gunners to stop shooting across the road and river and allow us to pass. I imagined horse-mounted knights waiting on a drawbridge to cart their wounded into the castle hospital. Passing EOD's truck felt particularly foreboding. It was laying on the passenger side, and the engine compartment was mangled and barely attached to the truck body. Three soldiers from Bronco security element and the EOD First Sergeant were laid over the top of the vehicle using it as a gunner's nest. EOD First Sergeant was picking off bad guys across the river using his Barrett M107 sniper rifle. He was not all there at the moment and could've used medical attention, but nobody tells a still "fit to fight" First Sergeant what to do. Bailey and Bigby were both outside of the overturned truck lying on the ground, being worked on by a couple Bronco medics and Pruitt. Bigby didn't look too bad but definitely couldn't continue to fight.

Bailey was in a little more trouble. Not life-threatening, but a good portion of his lower face had been smashed and ripped apart when he collided with equipment inside the vehicle. He needed serious and immediate medical attention. We unloaded Orton near Bailey and Bigby's triage point. When I got out and saw Orton again, he was nearly naked, unconscious, and had lost a fair amount of blood. His head was adorned with a dressing that had begun bleeding through, and his skin had a waxy or paste-like appearance. Hartzell put me on "keep him awake duty" and told me to "do a sternum rub every time he drifts off." This is where you rub your knuckles into someone's sternum so hard they wake up. It is taught in basic life-saving and first aid courses. What is not taught is that the receiver of your sternum rub may become violently alert. Orton's eyes would only focus for a few seconds before glazing over and rolling around like a marble on an uneven surface. I was rubbing his sternum frequently and it had begun to lose its effect.

Without consulting any of the half dozen medics working around me, I stuck my index and middle fingers as far under his clavicle as they'd go and gave Orton a good, solid tug. That woke him up. His eyes opened wide with anger and as soon as his gaze met mine, he punched me square in the eye while yelling, "I can't hear anything!" The effort sent him immediately back to sleep. He'd just been shot in the head, so I let it slide.

Around twenty minutes went by as we stabilized Orton and Bailey to be secured to litters for transport before we got the "five minutes until the bird arrives" warning. Once that call went over the net, we began carrying the two litter-bound casualties down to a farm field near a wide, flat spot on the riverbank a few hundred feet off the road. Bigby walked under his own power. Sutherland, Hartzell, and Pruitt carried Bailey while Ballantyne, Conn, and I carried Orton. At first glance, the field appeared to be flat terrain and sat maybe two feet lower than the road. In actuality it was several terraced fields of varying depth, but the grass was all cut to the same height. With two guys carrying the front of the litter and one in back, we tiptoed through the field to feel when the terrain would drop away. The litter still ended up breaking during our fall into the third terrace. With bullets still buzzing, we used our rifles as a makeshift brace to fix the litter and got moving again. By this point my legs were screaming. It was only a few hundred feet from the roadside to the LZ, but the terrain and terror I was feeling were murder on my muscles and ligaments.

About the time we finally arrived at the LZ, we could see the helo coming up through the canyon. Once the bird was a few feet from landing, a massive increase in enemy fire erupted from a new location across the river. The chopper was so close I probably could have stretched out my arms, jumped, and been able to touch it. But incoming fire was so intense the helicopter never got the chance to land and was instead forced to retreat. We were left out in the field with completely unprotected casualties. Our enemies were out of the reach of our rifles. The gun trucks up on the road, however, had plenty of capable weaponry. All we could do on the ground was make sure the casualties didn't sustain additional injuries. So, like ugly, human blankets, Ballantyne, Conn, and I draped ourselves over Orton to protect him from gunfire while Sutherland, Hartzell, and Pruitt covered down on Bailey. We weren't pretty, butts and faces everywhere. It wasn't really laying either. We propped ourselves over the top of the casualties, like a roof overhead. In the ensuing minutes, every muscle fiber in my body caught fire. While I wasn't particularly thrilled by our current situation, Conn was on his first big mission ever and was completely terrified.

With a sea of fear in his unbelievably wide eyes, Conn looked to Ballantyne and screamed "We're gonna die here!"

Ballantyne glanced over to me with an ear-to-ear grin made up of as much fear as courage, and valiantly proclaimed, "Yes, we are!"

He and I then high-fived each other over Conn's head and waited to be shot to death. For some reason, that ridiculous response calmed him right down. Even with the bullet quakes. You know how you can feel when someone nearby stomps the ground real hard? Well, we could absolutely feel lead impacting the ground around us and it was particularly unsettling. Dirt and grass bits splashed our uniforms and faces with unbelievable near misses. Days later, I even found a bullet lodged in the left heel of my boot.

After what seemed like eternity, the gun trucks on the road had done their job well enough for the bird to come back in and land. As it touched down, I was shocked to see how many fresh bullet holes adorned the hull. We quickly exchanged pertinent info with the flight medics and loaded the boys onto the chopper. As the MEDEVAC sprang back up into the sky, I felt ten tons lighter. Though bullets were still being exchanged, the five of us walked back to the road so slow it appeared as though we were elderly friends on a lazy afternoon stroll through a park. In truth, we simply could no longer muster the energy to move any faster. We reached the trucks, returned to the front of the platoon, and waited for recovery of EOD's truck to finish. Once complete, we resumed mission and made it to FOB Bostick without further incident. The trip home was also fairly low-key; I don't believe we took any fire on the way down.

Once again, the day after we returned to Fenty was a maintenance day and I had SCMO duty. I didn't have to, but for some reason I couldn't bring myself to turn over gear with pieces of my friend's flesh and bits of skull clinging to it. I went out back behind our hooch with a few water bottles and a new boot brush. I rinsed out his helmet and used it as a wash basin for the armor. It may have been done as some kind of honor to my friend, but all cleaning Orton's gear accomplished was to unravel me. I worried that we did irreparable damage when we dropped him, or that the way we performed emergency care was wrong. Minutes passed in my dazed yet racing state of mind and I didn't notice that blood had soaked my uniform up to my elbows and down through my pants onto my thighs. Due to emergency casualty care and cleaning its aftermath, I've worn the blood of my wounded brethren and best friends more times than I'm able to tally in one sitting. Not that the number is excessively high—it simply hurts too much to get through the few handfuls.

Silently, tears began steadily falling from my chin as I scrubbed the copper-splotched armor. Images, sounds, and aromas from the previous day swirled and intermixed with the task I was actually seeing and doing at the moment. When I instinctively wiped to clear my vision up, I smeared some blood and stringy flesh onto my face. That brought rage into my overcrowded head and upset me more. I was no longer silent. Nelson's room was at the back of our hooch, and he had heard the outburst of anger and anguish. From the back door he peeked outside, decided he was not going to step into my line of mental fire, and went out the front to find First Squad's leader, Staff Sergeant Jones. Once located, Nelson informed him that I was in need of some immediate help. By the time Jones appeared, I had knocked over the wash basin and devolved into a full-blown, babbling slump of blood and tears. Through a pile of panicky expletives, Jones hoisted me up, walked me to the shower point, and tossed me into a stall, clothes and all. He talked me down while the water washed away the blood and cooled my searing nerves. After the shower, we walked back to the hooches, I changed uniforms, and finished cleaning Orton's gear. Administratively, the rest of inventorying went quickly and smoothly. Emotionally, it was a waking nightmare. I wondered who I was going to have to inventory next. *Would they be injured, or dead? Would somebody be inventorying my gear?*

Almost nobody had seen me lose it, but a platoon is a tight-knit thing. Everyone knew. Instead of being made fun of or ostracized for falling apart, I was privately approached by almost everyone in the platoon. They came to check in and let me know I wasn't alone. I kept hearing how the others were as worried as I was. How they were afraid it would be worse the next time someone got hurt. *How long could we stay this lucky?* How they felt rage and fear boiling in equal parts like I did. Some even asked about how I was able to put myself back together so quickly after losing my cool. I don't know what bullshit answer I came up with back then but truthfully, it was them: my soldiers looking to me as a source of strength was what I needed to force me back to work and continue forward. I hadn't kept my promise about Lissy and Orton going home safe, but they were going to live. There were still plenty guys I needed to keep my shit together for.

Chapter 7

Bodies Just Plain Break

Whether a large detonation destroys your vehicle and sends you to a hospital, or a small pop barely dents the armor, feeling an explosion up close and personal hurts in a very unique way. While I can't fully put it into words, imagine a top-level boxer punching every square inch of your body at the exact same instant. The kicker is, every punch feels as though it originates from inside your own body.

Ever-advancing armor technology on our vehicles and bodies has made surviving large detonations of high explosives much more possible than in previous wars and conflicts. Body parts and truck panels remain in place and attached, but the blast wave itself still travels right on through the giant bag of bones and goo we call our bodies. Organs feel as though they've been battered and somehow shoved into the wrong location. Confusion, inexplicable pain, and vomiting are commonplace. Whether a strong or weak blast, an intense headache always comes with it. Motor function and speech sometimes don't quite work right for a while. Sometimes some things never quite work right again. In any case, it's a hell of a thing to be forced airborne by a bomb. There might have been more, but I recall six occasions where explosives struck my vehicle during my tour in Afghanistan.

The first was a low-order detonation on our way to FOB Connolly early on in the tour. A low-order detonation is where a small percentage of the explosive detonates and the rest fails. This causes the undetonated explosive material to be projected at the vehicle like a mud pie. We stopped, did a post-blast analysis for our report, scraped the failed goop off the truck, and away we went.

The second explosive to strike my truck was more substantial, but negligible, nonetheless. On our first or second BOW, a few RPGs were shot at us with most hitting the ground between our rollers and front bumper. One did happen to damage our right front wheel. With no soldier injuries, we limped to an outpost, changed the tire and wheel, and finished the mission. The final four hits occurred in a span of twenty-eight days beginning less than a week after Orton was shot.

20 April 2011

We were just finishing up our first (and my only) night mission. Karzai's Twelve, the rules put down by President Karzai that US troops had to abide by, specifically prohibited night missions unless special circumstances arose. We were given one opportunity to test the viability of route clearance at night. We had argued that it was more dangerous to find an IED the hard way, during the day when towns are full of people going about their daily lives, than when the streets are empty at night. I can't remember if we found one or not, but we ended the mission successfully and had turned back onto the main highway leading back to FOB Fenty. Twenty minutes after the sun had risen, a white sedan jumped lanes about two hundred meters in front of us. This was odd because VBIEDs (Vehicle Borne Improvised Explosive Device, pronounced vee beds) don't usually start acting squirrelly until they are much closer to their target, lest they give themselves away as an enemy combatant. But there he was to my bewilderment, just doing his own crazed version of a serpentine charge right at us. He slammed into a concrete divider but, to our amazement, kept his car in motion. We all assumed he wasn't a VBIED, just someone who had lost control of his car and was about to crash. As the perilously moving vehicle wobbled within fifty meters of our rig, we got our gear ready to aid the impending crash victim. Fitz, one of my vulnerable point search dismounts, laughed aloud.

"Dude wants to play chicken usin' a Toyota? Bwahahahaha! He's gonna splat the shit out of—"

In a spectacular ball of fire, car and body parts assaulted the front of my vehicle while a moderately-sized blast wave battered my stupefied face. We had a joystick-controlled gun turret that day, so my entire crew was safely buttoned up when the vehicle detonated. There was shit everywhere. The guy's license plate landed three vehicles back. My guys checked each other out for injury while I filled in the rest of the platoon over the radio. Our truck sustained almost no damage, so Sutherland moved us forward of the chaos to cordon off the street and block passersby. Fitz, Ballantyne, and I got out to start the post-blast analysis and damage control. There had been an elderly local on a bicycle nearby who sustained shrapnel wounds to his legs and back. We called in a MEDEVAC for him, but since he was not a US soldier and there were no US casualties, we were declined. Brummett, our platoon sergeant, had the medic, so he brought up his truck and team.

Marsh was our medic this run, and he got right to work on the old man. When he was through stitching up bicycle guy, he performed concussion evaluation tests on my whole crew. This is a quick series of questions medics ask after a blast to check for brain damage.

"What is today's date? Who are you? Who is the president? Where are you right now?"

That sort of thing. Along with the questions, blast victims are required to remember five words. At the beginning of the test, the five words are stated and then repeated back to the medic. Then the questions are asked, and physical checks conducted. At the end, all we had to do was recall those five words. The words were always chosen from three lists. As part of getting ready for the day, my crew routinely practiced memorizing all three separate word sets. I still remember one of them: Elbow, Apple, Bubble, Saddle, Carpet. We easily passed and continued mission on home to Fenty.

24 April 2011

The next slap came four days later on Easter. We were in the boonies again, working for FOB Connolly out near some remote farms. The roads were completely dirt, and the going was extremely slow. Each farm consisted of vast, open fields of agriculture or poppy, and one big farmhouse. Farmhouses were surrounded by ten-foot-high mud walls and a single door, no windows. We didn't see anyone anywhere. Not in the fields, not working the roadside farm stands, nothing. Unpopulated social hubs are a huge indicator something bad is about to happen.

It was nearing midday. We were passing yet another farmhouse when our engine and the entire compartment housing was blown off of our vehicle. I don't remember it super clearly, but I woke up vomiting. My face had broken the screen to my GPS and I couldn't see out of my left eye. Nothing was obviously wrong; it just wasn't working at the moment. Nobody was hurt too bad, and we were all able to exit the vehicle under our own leg power. Recovery of the vehicle took a while, but we were carted off to see the medic. After our evaluation test was given, it was determined that we weren't urgent—the worst of us had minor scrapes and mild concussions—and could just ride in the back of the CASEVAC truck, no MEDEVAC necessary. This ended up being a bad idea. Our truck was destroyed and since it was in two very big pieces, it was beyond the capabilities of the wrecker to tow it back home. Hell, part of the truck was blown over the walls of a nearby farmhouse. Brummett knocked on the farmhouse door for a few minutes, but no one answered. Our new EOD guy put a breaching charge on the door and blew the hinges so a couple Joes could grab the truck part.

To recover our truck, a team all the way back at FOB Fenty was spun up to drive a flatbed semi-truck called a "916" out to us. A couple of hours later and the recovery team had arrived, gotten the truck loaded, and we were all on our way back to Fenty. Less than a kilometer before we would have made it back to the actual paved road, the 916 driver cut a turn too tight and rolled slightly down a mountain. Idiotically, he wasn't wearing his helmet or seatbelt when it happened, and he hit his head a few times. This caused us to have to call in a

MEDEVAC chopper for the recovery team. We ended up stuck in the back of the crowded truck for about forty hours while re-recovering my truck and numb nuts got a free flight back to base. Three of the five of us had to take a day off before returning to mission. The rest of April was uneventful.

9 May 2011

The ninth of May found us on another BOW. We had already passed the 883 grid line and were way up in the mountain pass where the road is barely a single lane. On our left was a sheer rock face so tall we couldn't see the sky. On our right was a quaint little adobe village whose inhabitants specialized in terraced farming. Just past the village was the Kunar itself. Directly in front of us was a bomb. It wasn't in plain sight or anything, but it was obvious we arrived before they finished putting it in the ground. There was a small, one-inch wide, maybe one-inch deep, dirt trench stretching across the road starting from the village side and ending a foot before the rock face. At the rock face side was a big pile of freshly disturbed earth. It might as well have had one of those old neon signs with an arrow pointing directly at it. We talked it over in the truck. We knew what it was but there was nowhere to go. The road wasn't wide enough to turn around and the idea of backing up, possibly for miles, along this road simply wasn't plausible. Our only option was to try and skirt alongside the IED and get past it to allow the Buffalo team to come up and interrogate. No one in my truck was pleased by this, so we called it up to let the platoon know what was going on and get some more input. After a quick discussion with the other trucks, the decision to pass was made. We buttoned up the truck just in case and with teeth clenched, eyes open, butt cheeks squeezed, off we went.

"I do not feel comfortable with this," was all Sutherland said in protest.

As our rollers passed where we suspected the main charge to be, it detonated. The majority of the blast went between the rollers and our front end doing little damage, but the flames were insane. The IED had a secondary accelerant designed to make a big fire at the time of detonation and this pyrotechnic show occurred right on our hood. It wasn't normal camping fire either— it was solid red, torch-like flames straight from the bowels of hell itself. For the better part of a minute, all you could see out of our windshield was one giant, red flame and it was very hot. I came out of my trance when radio traffic about my truck passed into my ear. Stoof, the driver of the vehicle behind us, had witnessed the whole thing and was relaying information to the rest of the platoon. I broke in to stop anyone from flying on up to save us.

"Break. Break. Break. This is One-Three. We are okay and not on fire. It's just accelerant burning off."

Sutherland looked at me, eyes wide with incredulity, and very calmly asked, "And what exactly the fuck would you call that, Sergeant?" pointing at our window.

I grimaced at him, and with my hand over the mic asked Brummett for some time.

"Give us a sec and we'll put it out. I fucking hate IEDs." He chuckled to show that he was relieved we weren't crispy critters. Brummett chose not to send trucks up. To be honest, the fire burned itself out completely only a few seconds after the radio chatter. We sustained no major injuries.

Super EOD who breached the farmhouse was with us for this BOW and came up to do the post-blast analysis. He was unfamiliar with the operation and did not understand the full risk of being dismounted in the area. But after the blast, EOD temporarily seized command of the group. This is routine and no big deal. We found the command wire used to detonate the IED, so he took Fitz and Bosen and followed the command wire to locate the trigger point near the river. After a couple hundred meters of walking, my guys began to protest that they felt it was a trap and that they were too far from the cover of the gun trucks. Super EOD decided that my boys' opinions didn't matter and continued damn near to the water's edge, where they were promptly ambushed from across the river with heavy weapons fire. My kids were now pinned down in a grass field, hundreds of meters from the route clearance package. The ensuing radio traffic was almost entirely Fitz and Bosen scream-coordinating. My boys weren't injured, but it was loud, adrenaline was extremely high, and minutes earlier they had been in a vehicle explosion. My kids needed to keep their heads as cool as possible, so I put on my best radio guy voice and keyed the mic.

"This is One-Three. Stay low and in cover boys, we're on our way."

One of the boys tossed smoke twenty meters east of their position to aid with retrieval. Without a word between us, Sutherland backed us up a few feet and immediately drove straight down a hill entirely unsuitable for driving while Heiner dialed the fifty in on the enemies shooting at our brothers. BOW would not take from us a third time. The trucks in back all hastily maneuvered into positions where they could provide cover fire to the dismounts and relay positions to us. We could no longer see any roads, just fields, houses, walls, and debris. It was an extremely rough two-minute drive. Sutherland located—okay, landed on—a small side road that led directly to the fields our dismounts were pinned down in. Heiner unleashed lead rain. With accurate suppressive and destructive fire now pouring out of all of our trucks, the dismounts could once again begin moving. Shit doesn't go this coordinated or well-choreographed unless it's Hollywood. I'd never seen anything like it. Our accuracy, communication, and calls for fire were all on point. Nobody was gonna let BOW win again. Within mere minutes, it was all over without any major injuries and my boys were back in the rig.

Later in the year, Super EOD would get his by being shot in the thigh during another post-blast analysis on another BOW because he decided to flip off a mountain instead of take cover. So even though I never did get the opportunity to give that senior-ranking, arrogant fool a piece of my mind, it all ended well. We made it to Bostick without any further incident and got checked out by Bostick's medical staff. With our well-rehearsed responses, we were cleared to drive back to FOB Fenty the next morning. But things were beginning to negatively add up for me. My words and thoughts came slower and cloudier. Body parts began to ache and crack. My hands stopped being steady. Most importantly, I forgot more than I remembered in the day-to-day.

17 May 2011

Lo and behold, we were operating for FOB Connolly out near those big farms again. It was early afternoon, our mission set was already complete, and we were actually clearing our way back to the FOB. Our Husky operator had been hit pretty hard earlier in the morning, so the wrecker already had a vehicle on it and the CASEVAC truck was one passenger heavy. We were maybe an hour outside of Connolly when we came across a burned-out Russian tank that had been sitting in the exact same spot since the end of the Soviet-Afghan War in 1989. We'd seen it dozens of times before and it had become a kind of "Almost there!" marker for missions out in FOB Connolly's area of operation. We always made some remarks as we passed: "Poor Russians." "Can't believe that was a tank." "That area was not clear," etc. However, there wasn't much sarcasm or arrogance with the joking as we knew what getting hit was like. We were on edge.

It was spring and a small irrigation stream had been cut and filled along the left side of the road. No more than thirty inches wide and barely a few inches deep, the water was clearly used to irrigate only the two fields butting up on either side of the road. Each field was bordered by small adobe walls, maybe two feet tall and a foot wide, stretching the entire length of the field. Every few dozen feet there were smaller streams jutting through the walls and into the field on the left or across the road to the field on the right. These smaller streams across the road were only an inch or two wide and less than that in depth. Just big enough to constantly trickle water to the other side. The stream was clearly a seasonal occurrence, as a large tree had been able to grow to a nice mature size along the road directly where the stream entered the left field. This was an area where trees didn't perform all that well, let alone grow large canopies. Even though we all had varying degrees of the heebie-jeebies, there was enough evidence to call the area clear. Ten or so meters past all the alarming stuff, we got pounded so hard I don't remember the hit at all.

I woke up to that same, brown-colored world of pain I had visited years earlier in Albu Bali near that IED factory. Every point on my body was screaming, my left foot wasn't working properly, and I couldn't see out of my left eye again. I must have slipped back out of it because the next thing I recall is Sutherland yelling at me to wake up and open my door as his was busted and wouldn't budge. A strange, hazy déjà vu moment occurred when I turned around and saw Heiner, Bosen, and Fitz in back trying to get the rear door open, but the blast had broken the combat lock. Fitz began bashing the back door with a hatchet to no avail. Heiner suggested they go out the top and out they went.

From what I could tell, none of us was in pristine shape but nobody was in life-threatening danger either, so I radioed to tell everyone we were fine, but the radio was definitely broken. After the initial blast destroyed the engine, a fire had broken out in what was left of the compartment. Nothing worked. It didn't really matter as no discernible words would come out of my mouth when I spoke. In the midst of the chaos, I remember thinking, *That's a lot of gray and orange on my window.* Sutherland grabbed me by the armor and shook me pretty good. "Sergeant, stay awake and open the fucking door, okay?!" I looked at him triumphantly and vomited on myself while struggling to get the no-longer-power-assisted door open. The door eventually relented. Sutherland and I popped out of the truck minutes before it became engulfed in flames.

Things got even fuzzier for me after the door opened. I sat by the right wall trying to stop spitting up on myself for a while, in and out of an uncontrollable nap time. Then someone helped us walk over into the left field to wait for MEDEVAC. The helicopter landed and took all five of us directly to FOB Fenty for medical aid. I remember the ride being strangely peaceful. My left eye was working again, but still blurry and the brown had almost faded back out of my vision. The helicopter door was left open, or maybe it didn't have a door. But the view was nice. Endless little farmhouses scattered amongst little green fields. Occasionally the giant black plume emanating from my truck could be seen. To my irony-filled dismay, that burned-out Russian tank wasn't even a block away from the raging bonfire formerly known as my truck. I'm still bitter about all of my personal shit going up in flames, but I like being alive more than having stuff. Days later the boys sifted through the wreckage and among the things recovered was the now-carbonized knife sharpener Bull had given me years earlier. As it was the last piece of him I had, it meant the world to me that the boys had found it. The rest was replaceable.

Back on the ground, because of some rockets, grenades, flares, and an exceptionally large stockpile of ammunition in my truck, no one could get close enough to put the fire out. Not that anyone had any fire extinguishers left. The priority had been on containing the fire long enough to get soldiers out. Gear was secondary. There just wasn't that much time. It was a fuel fire in an engine that used jet propellant #8. The Halon fire suppression system even went off. In the few minutes it took to get my crew safely aboard the helicopter, the fire had enveloped the entire truck. All that ordnance cooked off for several hours before recovery of the smoldering truck could happen.

At Fenty, Sutherland and I were put on oxygen while Bosen, Heiner, and Fitz were evaluated in another room. We were both laying on gurneys, shirtless with clear masks strapped to our faces, hooked up to myriad machines that constantly took readings of our vitals. People kept coming in and talking to us with words. I was certain they were words; I just couldn't quite grasp them. I'd just stare at the people and every once in a while a word here or there would make sense. I'd blurb out an answer I thought they'd want followed by a proclamation of how okay I was. No one believed me this time. However, when a high-ranking sergeant major whose name I can't recall stepped in to see how the injured troops who lived on his base were doing, his reaction caused the entire room—the medical personnel, me, and anyone else within earshot—to burst into uncontrollable laughter.

"What the fuck are tho... are those nipple rings? I don't need to see that shit!" And out the door he went. Never spoke a word to us. One of the best "high-ranking stranger wants to check in on injured soldiers they have never met before because it is routine and mildly obligatory" moment of my military career. They're usually overlong and more awkward.

That was the last funny thing that happened to me for a while. Bosen, Heiner, and Sutherland were released from the hospital late that night. They'd be on restricted duty for a while and, if they passed their medical reevaluations in a few days, they'd be back in business.

Bosen and Heiner got back to work relatively quickly. Sutherland took a bit longer, but none of them were quite right afterwards. Fitz and I were sent to the on-base TBI clinic for a "few days of in-person monitoring."

While there, we were to be monitored by a Dr. Someone and Sergeant Greniger. I hated his guts instantly because he was peppy and smiled all the time. He was also quick-witted. Peter Taillac had given me a shaving profile (permission to not shave every day and keep a one-eighth-inch beard) because I hated shaving. Greniger bested me in a battle of wits over something and nicknamed me "The Bearded Lady." Essentially, Dr. Someone did doctor things while Greniger babysat the clinic tenants all day. He ceaselessly crushed us at board games, and we seasoned everything he ate with expired chewing tobacco. It was give-and-take. We'd spent a few days there, and he clearly was a guy who cared. Getting us engaged was his thing. When the report from Dr. Someone came down that we were to be sent to Bagram Airfield (BAF) for further medical care, his demeanor gave it all away. He wasn't arrogant, loud, or happy.

"Guys, you're trying. You suck at everything, but you're trying. If it were up to me, that's all you'd need. But neither of you is getting better and you both need to go to a real hospital. You'll be flown to BAF tonight. Go to your hooches, get clothes, whatever you need, and say your goodbyes. Be here after dinner chow. Sorry, boys." It was over and I knew it. Not exactly knew, but everything felt different about that flight.

When we got to the BAF TBI clinic, there were multiple buildings a few feet from an actual hospital with concrete walls, with real hallways, and everything else. Instead of a staff of two, the BAF clinic had near ten soldiers on babysitting shifts, but no doctors. Our appointments were all in the main hospital and it was so close that no live-in doc was needed. Door-to-door walking from clinic to hospital was maybe two minutes.

Fitz and I spent our mornings and afternoons going to see this doctor or that one, whom we lied to as often as possible to expedite going back to Fenty. Every evening was a test or training for it. The test consisted of riding a bicycle for five or ten minutes without falling off, passing out, or throwing up. That's it. Day in and day out, we failed. Sometimes Fitz threw up and I fell off. Sometimes Fitz fell off and I passed out. We never did sync up. The bikes were set up in a line of about six, and we always took bikes right next to each other. Though we never accomplished it, we joked that we'd pass the test by having seizures and falling into each other instead of falling off. At night we'd plot completely unreasonable ways of rigging the bikes to hold us up, or how to make time go by faster during the test so twelve seconds appeared to be ten full minutes to the stopwatch.

I also thought a lot about a guy from my Ramadi tour, Groomer. We called him Bones on account of that's what he was, a bag of ribs and elbows. He was nineteen years old, five foot ten-ish, short, dark hair, glasses, and weighed perhaps one hundred twenty pounds soaking wet. Early on in the tour, his truck was hit and flipped onto its side. The gas tank had ruptured with fuel and oil leaking all over him when he came to. The blast had bent the steering wheel, column, and part of the seat frame, pinning him in place. His squad leader's leg had been caught between the radio mount and the armor wall, and though still attached, it had been snapped in two with blood spewing everywhere. As if someone else's blood adding itself to the list of liquids pooling on Groomer's face and body wasn't enough, an engine fire broke out and threatened to burn him to death. Thankfully our CASEVAC element got to them quick enough to put out the fire and get everyone out of the truck. Groomer had still been trapped inside that truck for the better part of five minutes just staring at a small flame hoping it didn't grow.

I helped clean the truck out later that night and it wasn't pretty. Large quantities of blood emit the pungent aroma of a wet dog made of iron. Psychologically speaking, we all would've understood if Groomer had decided to punch out and go home on medical. However, his injuries weren't serious enough for the army to make him go home and after a few days of rest and recuperation, he was chomping at the bit to get back out there and join the platoon. He was allowed to take a simple physical test every week to see if he was ready to go back on mission. I don't recall his exact version, but it was something standard like, "Hold so much weight and walk this far while wearing armor, no passing out or vomiting." He failed miserably every single week for a long time. Inevitably he'd be told something along the lines of, "I'm sorry, Bones, not this time. We'll try again next week." He spent over two months working on base, fixing trucks, taking care of admin jobs, and testing once a week. Eventually he did pass and returned to missions with the platoon. But the time he was away were tough missions and he always felt bad about not being able to be there with us. Years later, he told me this over the phone. I felt compelled to tell him my side of the story.

I saw a man fail day in and day out to get through simple soldiering tasks and rejoin his brothers. I watched for dozens of meals as he struggled to chew his food because eating caused so much pain in his head. Unbelievably, I watched him lose more weight and feebly battle to carry out simple labor tasks that I had no problems doing. However, instead of seeing someone weak or damaged, I saw a fighting man slowly getting his legs back underneath him, and that inspired the whole platoon to push themselves further. I specifically remember getting a water bottle out of the community fridge one afternoon before mission and feeling a click and some pain shooting down my left arm. Standing by the fridge and rubbing my tingly, semi-functional arm, I thought, *This is starting to be a problem. Maybe I can get someone to drive for me tonight*. But up came Groomer in his PT uniform, covered in dirt and sweat.

"Hey, Bones, looks like you tested. You rolling tonight?"

"I was so close today, tossed cookies right near the end. I'm definitely getting there, though!" His response was optimistic and determined in tone.

We talked for a bit more and as I walked to my hooch to put on my uniform, my shoulder didn't seem to hurt so bad. Fast-forward to my stay at the BAF clinic; my respect for Groomer found a whole new depth.

Eventually Fitz and I did pass. After some really hard conversations with doctors and our company commander, we were allowed back to Fenty, but with heavy restrictions in place. All in all we got back to Fenty by mid-June. Fitz could only go out on missions that were deemed a milk run or easier. On his first dismount operation though, he collapsed and after being evaluated back at base, was sent all the way to Fort Carson, Colorado for further care. His deployment and career were over. I remained in-country but was barred from leaving the base unless the commander gave his say-so, which didn't happen often. Captain Sparrow had been in Ramadi and knew my blast history well. He admitted that my presence was a morale benefit to the welfare of the men. But he stated in no uncertain terms that I was no longer mission capable and wouldn't be rolling. It was a gut shot to hear, but it was true.

It was obvious I wasn't completely mentally present anymore. To cover up that reading had become nearly impossible. I had each of my boys read their part of the mission brief out loud because it would "enhance everyone's battlefield awareness." Hell, one of my own kids nicknamed me "Guacamole Brain" because I lost things so often. I had four or five guys looking for my keys for over an hour once. Leavitt found them, in the lock, hanging on my hooch door latch. On a rare occasion that I did get to leave the wire, I was put in charge of the stay-back crew on Connolly where I crash-landed a $35,000 observation drone called a Raven directly into Connolly's burn pit. Thanks to Stoof running into the pit and grabbing what he could before melting his boots and scorching his uniform, some pieces of it were saved. I wasn't allowed out after that. Every non-combat/base-bound duty that came up was mine, but it didn't bother me so long as I remained useful.

The definitive blow to my ego was when I was fired from squad leader. Vargas, who had been an amazing alpha team leader, vehicle operator, and friend to me would be taking over Third. There was no one better to replace me, but it hurt so fucking much to realize that, above everything else, I was also no longer capable of leading my guys. After a few months of struggling to keep it together on a basic soldiering level, the tour ended, and we all went home. Sort of.

We arrived at Fort Dix, New Jersey in late September for Demobilization processing (DeMOB) and our wounded from earlier in the tour were almost all there to greet us. Seeing them made everyone's mood better. Seeing Orton and Lissy walking, laughing, and talking was a sight I will never forget. For the first time in what felt like forever, I could relax and breathe. We got the rest of the night to catch up and visit. The next morning it was time to pay up. The big thing I had promised Sparrow was that I wouldn't lie once we were home in the States. When deployments end, we go to DeMOB where lots of doctors ask even more questions and this time, they'd have all the machines they'd need to check everything out. Lots of soldiers lie because if you admit that even the slightest thing is wrong, your trip home is instantly delayed by weeks or months while you get medically checked out. I did as promised: I spoke about what was hurting and the trouble with reading and my eyes not focusing quite right. It was a blur of tests and appointments going over the results. Images of my neck, back, legs, and arms were taken with and without contrast (that awful imaging dye they inject you with that makes you have to pee). When they came in, it was like an entire photo album of how bodies shouldn't look. Broken ribs that healed awkward here, a few fractured vertebrae there, crushed nerve clusters; every day another poorly healed-over injury was discovered, and I was beginning to understand that this wasn't a checkup: it was the end of the line for my soldiering. I began slipping into a depressive, anger-filled state. After three weeks, only the medical hold-overs remained. The rest of the company had gone home.

When the piles of tests and diagnosing were all finally completed, I was called into a meeting with my case manager. She told me that I needed to attend another TBI clinic for a few days. I wasn't thrilled but she said it would actually help me get home faster if I went. With that carrot dangling in front of me, I agreed to go, no questions asked. Early morning on the 2nd of November, a duty van picked me up and drove the one hundred sixty-three miles to Walter Reed National Military Medical Center in Bethesda, Maryland, where I would be attending their "live-in" TBI clinic.

Chapter 8

Suffering Is a Place

Walter Reed does not have a TBI clinic. What they do have is a full-scale, inpatient wing on the seventh floor of the main hospital building, and I was not happy to be a resident. As if I was a suicide watch, hospital staff took my tobacco and all my weapons, even my pocketknife. After being extremely hostile for three days, I was locked down in my room and the army called my mom back in Utah. Pro Tip: Do not break things with office chairs and threaten to impale hospital staff if you don't wish to be restrained.

The head doctor informed my mother of what was going on and how bad my injuries really were and asked whether she would be willing to leave her job for an unknown but likely extended period of time to come live with me as my Activities of Daily Living Assistant, or ADLA. Wherever I went, she would go, appointments and all. The main thing was they wanted someone familiar to me who could help me understand what was happening and who could also help the doctors understand me when I couldn't properly communicate what I needed to. Without my guys around to cover down when I passed out, forgot what I was doing, or needed an explanation because someone was telling me something I didn't understand, I quickly realized how poorly I was functioning, and it was beyond terrifying. I had no idea so much had been lost and I needed help coming to terms with it. Ever the mama bear, she was on a plane the next day.

I would not have made it out of Walter Reed alive if not for my mom. Once she arrived, things went a little smoother and I relaxed some. She developed relationships with the staff and convinced me to stop attacking/threatening them. In truth, I think I was so hostile because this wasn't a standard checkup or routine care. I knew this clinic was a full-blown hospital stay that was going to end my military career and I was angry about having zero control over my own life. I hadn't been much of a success story before the military and now I was mentally deficient, mildly broken, and being fired from the only thing I'd ever been good at. Days were filled with doctors' appointments galore, speech pathology, MRIs, X-rays, physical therapy, brain therapy, a small procedure here or there, and neurology and all the other -ologies for busted parts. Turns out I'd had more than my share of broken bones, nerve damage, and joint dislocations. Not everything healed correctly.

It went on and on; every day a new problem I had ignored for years or months showed up to hammer nails in the coffin of my career and I plunged deeper into a personal despair. There were many days where I just could not keep it together mentally. On one such day I was outside one of the main hospital buildings sitting on a concrete bench. The bench was in an ambulance/patient loading/unloading circle driveway area. To church it up, the area was pleasantly landscaped with ornamental trees and bushes lining the sidewalk, a handful of two-seater concrete benches, a dark-tinted-glass bus-stop-looking smoking area, and a big, beautiful, green lawn off to the side. I had been on the farthest bench from the doors for a good twenty minutes crying silently. My mom was inside the hospital raising hell about I don't know what, but she was mad, and my being upset was the cause. Honestly, I just remember staring at the big, red "no parking" paint line on the curb through blurry eyes and feeling incredibly shitty about my existence.

My weeping was apparently not as quiet as I thought. Amid my woe-is-me fest, a very faint and gentle hand appeared on my right shoulder. His touch was so light I really only felt his pinkie on the base of my neck. At the same time I noticed the pinkie, a gruff but kindly reassuring voice emanated from behind me.

"I don't know what it is and I don't know what happened. I'm sorry. It gets better if you keep fighting, man. Promise."

Without another word, his hand lifted off my shoulder and he walked back toward the smoking area. I don't know how long I sat there, but I stewed angrily for several seething minutes. This guy didn't know me and had no right to stick his nose in my business. Anger continued building as I slowly got up to turn around and confront this interloper of my emotional breakdown. However, the moment my gaze landed upon him I was rendered dumbstruck and mute. The man was wearing a physical training uniform drenched in sweat and he was on a smoke break. Both of his legs were equipped with the raptor-styled running prosthetic, one of his arms was a wire-and-claw-styled hook, and a lit cigarette was perched between the fingers of his only remaining flesh and blood appendage. He was laughing with his physical therapist about having become faster than ever.

Without saying anything at all, I walked inside to my mom, tugged on her arm, and we peacefully left for chow. Over the coming weeks I realized human suffering is a place—a town, if you will. Representing all the hardships life has to offer, everyone will eventually find themselves on the streets of Suffering. No one takes the same road in. No matter where you arrive, the only road out is all the way across town. There are no inns, taxis, or diners, just your entrance, a desolate and deserted middle, and your exit. Though you may come across other people on your trek through town, and some may offer words of encouragement, you can't save them and they can't save you. All you can do is see you aren't the only one in town and continue plodding your way out. No matter the walk of life, successes, or failures, we all visit Suffering from time to time, and that makes comrades of us all. Even though it didn't magically make my recovery process easier, seeing others struggle with the same monsters as I did made facing mine more bearable. Someone physically devastated by war took time during his own recovery to notice my anguish and console me. Someone who clearly had a more difficult street to walk down than me. I still can't thank him enough for helping me put things into perspective that day. Thank you, One-Armed Guy. You're yet another person who, knowingly or not, saved my life from spiraling into oblivion.

I met more and more people whose lives were as ripped apart as mine or worse but were further along in their healing journeys. Did a lot of puzzles in my spare time in the common room. There wasn't much to do in the Seventh Ward so by Thanksgiving, I was a puzzling pro, sort of. Another guest of the clinic who was also an avid puzzler was Nick, a kind fella whose left eye had been blown out of his head by a parachute grenade a few years earlier. Over those years his intermittent stays were always long, so he enjoyed the puzzling company. He didn't wear the same clothes often, but he wore one shirt at least a few times a week. It was from that movie *A Christmas Story* about a little boy named Ralphie and how he wants a Red Ryder BB gun for Christmas. Throughout the movie every time he brings it up, he gets told they're too dangerous and he could shoot out an eye. Well, his determination does get him the rifle on Christmas morning, and he goes right out back to give it a test. He also immediately shoots and breaks one of the lenses of his glasses. Great film. Nick's shirt had Ralphie's face wearing his broken glasses, the Red Ryder BB gun underneath his face, and a caption that read, "You'll shoot your eye out." It took me a while to put two and two together and realize he had added an iron-on arrow pointing straight up to his own eye. Here was a guy who had lost an eye in war and his favorite T-shirt poked fun at his own resulting disability. After that, Guacamole Brain sounded more endearing.

A month or so after my mom arrived, things had progressed to the point where I could move out of the TBI wing of the main hospital. My mom and I were both transferred to soldier recovery housing, a kind of two-bedroom apartment. The front door opened to a small living room with a kitchenette. Similar to a hotel layout, the living room was furnished with a couch, a TV on the wall in front of it, and a small desk and lamp. The kitchen section was an island counter with a couple bar stools, a fridge, stove, microwave, and some cupboards all tucked together against the wall to the left of the front door. The bedrooms were on opposite sides of the living room, each having their own bathroom and shower. Instead of heading to another building, ascending seven levels, and getting past a floor-to-ceiling, tempered-glass security wall with keypad entry to come see me in the TBI ward, my mom just had to stroll across a room. That made life a lot easier. I felt more independent and positive about eventually being able to take care of myself. Plus living in a locked-down hospital wing never sat well with me.

With our newfound independence, one of the things we did for recreation was walk all over the hospital campus and the immediate surrounding area. I mean, we had to go on walks for some of my therapy, but we wandered the hospital grounds on several occasions just to be outside. Walter Reed is dotted with grass fields and tree groves intertwined with both old two- or three-story brick buildings and new, larger, modernized glass monoliths. Deer can occasionally be found foraging around picnic tables in these miniature nature parks. One day toward the end of our time there, my mom and I decided to take our walk off base and get some sushi. Christmas was closing in, and we wanted to celebrate how far we'd come in just over a month and a half. To leave Walter Reed on foot, there are walk gates on either side of the road leading out.

As we approached the left gate we noticed a large gaggle of geese loitering on the grass nearby. Not knowing that geese were anything but pretty, I walked on over to investigate. My mom chose to keep her distance and said nothing. Excited to observe these beautiful fowl, I giddily left the sidewalk and plodded onto the grass. The second I was within eight feet of the gaggle, a dozen or so of those sons of bitches broke off and charged angrily in my direction. Baffled and amused, I just stood there while they honked and flapped their way to me. As far as I was aware, geese were just big, harmless birds. Thirty seconds of being repeatedly beak-bonked about my legs, back, and ass, however, I was well-schooled. I didn't want to hurt the crotchety little bastards, so I slapped the most aggressive ones off me and speed-hobbled through the gate. Again showing her wisdom, my mom roared with laughter as she abandoned me, taking off across the street to the other gate. We met up safely outside and I had to question reality. Wide-eyed, I looked to my mom who was now walking over and yelled, "Did that shit really just happen? Was I just attacked by birds?"

My mom, tears running down her cheeks and still chuckling, replied, "Yes, baby. Geese are real and sometimes mean. It's fine, let's go get you some fish."

I glared at my mother with a "squinty eyes of scrutiny" kind of face, while she took a second to get her laughter under control. With birds and giggle fits defeated, we got our bearings and walked on to victory sushi.

When Christmas finally arrived, my mom and I were cleared to go home for a visit. By then, I was in a better place of accepting that my service was ending. We went on a present-shopping spree in the PX (sort of a military mini mall) and flew home for ten days. It was good to see everyone and while home, my mom made calls and started the ball rolling on releasing me to the Community-Based Warrior Transition Unit Utah (CBWTU). All of my major issues were now in the "managed" category, and the CBWTU was created specifically to allow soldiers to return to their home community to finish off any procedures, get permanent continuing care set up at their local VA, and transition back into civilian life. This is not a quick process because the army rule is simple: The army received me in a prime and healthy state. They will not release me until I am again in good working condition, or as close to it as possible. All told, I was home for good by the middle of January 2012, but I didn't retire until March 2014. Retirement would have gone faster if I had stayed at Walter Reed because all the medical facilities were centrally located within walking distance. The luxury of being close to my entire family and social support structure was worth the extra time.

During those two years I still had to report to the CBWTU every weekday, attend a never-ending appointment list, and get better. Occasionally charity groups would list events that soldiers in the CBWTU could attend to try and get reengaged with the community. Most of those programs were not for me, but a few stood out as seeming truly geared toward helping soldiers heal. To break the monotony of being the non-commissioned officer in charge of paper shredding in a medical unit, I did attend a few. Some organizations ended up having very strong impacts on my recovery, groups such as Warrior Bonfire Program, Warriors on the Cataract, and Warriors Afield Legacy Foundation (WALF). Orton called up one Wednesday afternoon in fall 2012 to let me know he had signed us up for a bird hunt/fishing thing called, "Cast and Blast," hosted by WALF.

"Hey, we're going on a thing Friday. It's three days. We fish and shoot stuff. I'll pick you up Friday morning. Captain Longley (our nurse case manager at the CBWTU) already approved it, you're going. Bye-bye."

Without giving me a chance to conjure excuses, he giddily hung up. I packed my bag, and Friday morning we drove to Castle Valley Outdoors, a ranch in Emery, Utah. Once we arrived, we were given our own rooms in a very nice lodge house and were each paired up with a Vietnam veteran who had also been injured in combat. It was three days of you and your partner walking in open fields or sitting on the banks of a pond just swapping stories with each other.

Even though our periods of service were nearly four decades apart, the ugly parts were eerily similar. My partner for the event was retired Army Colonel William "Bill" Hansen, who served in the sixties as a commander in the Eleventh Armored Cavalry Regiment in Vietnam. He recounted moments that were as awful as mine as if they were fond memories from his childhood. I couldn't believe how casually he brought up the things that had physically and emotionally demolished him. But over those three days, we talked less and less about combat and more and more about how he lives with it all.

Among his wisdom-filled words he said, "There isn't a magical trick to it and there won't be a day where you simply wake up burden-free. You just learn over time and accept, truly accept, that it's part of what makes you who you are. Once you accept it yourself, you won't have to keep everyone else so far away. You'll eventually be surrounded by people who understand and love you, faults and all." He leaned in close, smiled widely, and lowered his voice. "And once everyone knows you're nuts, you can say whatever the hell you want, and they'll still invite you to golf."

Hearing Bill's words was one thing; witnessing his decades-long relationships with his brothers firsthand was entirely different. Here were a handful of guys in their late sixties and early seventies who had gone to war when they were the same age as me. They had all made it to the other side of their traumas and won at life. Sure they had scars, limps, clicks, and ticks just like us, but they also had successful families, achieved huge life goals, and had years of learning to deal with it, each in their own way. They embraced the supposedly embarrassing moments where motor function or memory failed. They laughed loudly, recounting the antics of their fallen, never cowering when the laughter inevitably faded, bringing solemn, quiet reminders of loss. They understood and accepted it was their honor to feel that pain because somebody loved them enough to sacrifice. They were vehemently engaged in decades-long rivalries that frequently ended in mischief and mayhem.

Two retired generals with a long-standing marksmanship rivalry decided that the Cast and Blast, in the presence of all of us "young bucks," was the perfect time to settle old scores. Bill took some helpers outside and set a target about one hundred yards out from the lodge's back porch. Each general produced their own custom rifles and were allowed one shot at a time to hit the target. First to hit wins. A coin was flipped, and they began. I don't recall who went first or who even hit the damned thing.

All I can recall is the incessant badgering between the two men as they each missed their shot over and over, and the impressive fireball downrange when someone finally did smack it center mass. Bill had put a gas can and a small explosive device behind the bull's-eye of the target. It immediately caught a small patch of grass on fire. Orton, Kendall, and I all ran off for water. We resembled a poorly directed *Three Stooges* firemen one-act play, but with a lot more hobbling and foul language. We found some buckets, filled them, and took off out the back of the lodge. As soon as the three of us were all outside, we heard Bill's voice boom from behind us.

"What are you idiots doing?" He had a hand leaning on the deck railing holding a mixed drink and the other hand waving from left to right, as if to display the vast expanse of arid desert we were standing in.

"Where do you think that fire is planning to go? Come grab some burgers. It'll burn itself out."

As we walked past Bill and the other old guard members, they laughed warmly. Bill chuckled.

"Be sure to get a beer—you kids are high-strung."

The fire did, in fact, burn out before we had finished our meal. The rest of the weekend was full of similar antics and wisdom from Bill and the others. We still keep in touch and see each other a few times a year.

The months at CBWTU dragged on a lot less after the Cast and Blast and eventually March 2014 arrived. The last thing that held me up was going through the Medical Retirement Board process. It was in the middle of being revamped and way too convoluted for me to explain. In addition, I still don't totally understand it. I had received a 60% disability rating from the VA after Ramadi and figured they would just turn it back on. But when the time came, I was flown to Fort Carson, Colorado and was given the "retirement and you" package containing my Board-given rating. My gut was punched one final time. Several printed characters near the bottom of the last page finalized what I refused to mentally grasp. Now concreted in ink on legal paper, the words read: "100% disabled P&T." The last two letters stood for "permanent and total." It was done. I wasn't a soldier anymore. Bittersweet, that retirement was. I didn't know what to do. Sure, I was now free to do as I pleased, but the world didn't want or need me anymore. I had the paperwork to prove it. I was torn in two opposing directions that screamed different things about the same outcome. I felt worthless, but free to live in peace.

Adjusting to retirement was not as simple or freeing as I had envisioned. Therapy is ongoing to this day. It's more of a maintenance program or school for learning new techniques of self-control, than talking about trauma on a couch. But I have come to understand that from time to time my strength, both mental and physical, will falter, and getting assistance keeping my head on straight is something that's become my normal. I'm no longer embarrassed or self-critical about having to ask for that help. It's just something that occasionally needs doing. I have a chiropractor for when my spine flares up and waiting a few weeks for the VA is not a possibility, but for just about everything else I head over to George E. Wahlen VA Medical Center. Walter Reed was a bastion of the most up-to-date medical tech and procedures, but the VA in Salt Lake is near equally equipped and only thirteen miles from my home. I've been going off and on for fifteen years. I'm friends with some of the staff and other vets I've met, occasionally getting lucky and running into someone I served with. I'm thankful because even though I hate hospitals, it's an approachable and familiar place that helps me live successfully in the civilian world.

Problems with daily ticks and twitches exist. Forgetting words or stuttering is common; seizures are rarer. Some days my back reminds me that an army doctor actually described it once as looking like "a bag of smashed assholes." My limbs are not always completely under my command, but there are far more days of good mobility than days stuck feeling kinship with a disassembled crash test dummy. I can basically never go outside without dark lenses to protect my eyes, but sunglasses are perpetually cool, so who cares about that one. Days where I simply cannot get my brain to function properly are what we in the TBI world call "bad brain days." They are by far the worst thing I have to deal with physically. I can complain and bitch all I want about my legs not working right today, or my hands shaking too much to get any work done. But bad brain days, well, I can't tell you what goes wrong, but I can tell you what it's like. Imagine trying to tell a friend, "I'm going to the grocery store," but, "I'm tire iron," is all that comes out. It's like that, but for things like opening the fridge, driving a car, or stepping into the shower. All I can do is kick the side of the tub for five minutes while I figure out how to get into the damned thing. Complaining, or even explaining, to others when I don't understand it myself is next to impossible. So all I have to say, if anything at all, is "bad brain day," and those closest to me understand. Everyone else gets stuck watching a stranger make *derp doop* sounds while I figure out whatever is stumping me.

Over time, I've been continuously developing my own philosophy on what I should do or be. Way back in Basic Training we had an extra duty called area beautification. This is simply where you walk around the base and pick up garbage, making the area look pretty. It occurred to me that area beautification can be applied to all aspects of my life. All I have to do is try and leave everything better than when I first encountered it. Whether it be land, buildings, or my relationships with others, I believe it's worthy of attempt. It's a simple concept that, when applied to everything, becomes nearly impossible to achieve. And that is what I want the remainder of my life to be: one continuous attempt to do a better job of being a better human than I was able to achieve the previous day.

My late friend Moussa said it best: "A life lived well is one of servitude, not of gratitude."

I don't know if I interpreted his message correctly, but I got: "Giving for no reason at all is more rewarding than giving so someone can thank me." It's a work in progress. Continuing to seek out people with stories and wisdom beyond my own is how I'll refine it.

Obviously, I don't have everything or really anything figured out. What I know this day, and any other, is the direction I want to aim myself. I am certain to my core that I must continue through life attempting to live honorably until my bones and brain are so broken that I'm no longer of use. By through, I don't view life as a river with obstacles to be deftly navigated. I envision a single, monumental path to infinity, filled with walls. There's no way around or over. To move forward, each must be toppled. The only tool for demolition: a hammer comprised solely of willpower. Though my body is no longer as reliable as it once was, my will has grown to unfathomable depth thanks to all the people who have shown me what a human being is capable of enduring and accomplishing. I'm fully aware life isn't all kindness and pleasantries.

For the rest of my days, I'll occasionally find myself locked in combat with my demons. When my eyes close, burned men and children in pieces will plague my vision. Even after scrubbing, I'll sometimes feel awash with blood that isn't mine. Shame over missed bombs and the devastation wrought by them will try to creep in to embrace me. Hate and rage will scream from the depths to be fed even the smallest meal. It'll never fully be over, which is fine. It will mean I have honored my fallen by never giving in to the lies my heart and mind occasionally whisper to me. I understand that I will stumble over and over but will look forward to endlessly waging war against my darkness with a defiant smile on my face, nonetheless. The last decade alone has proved my mind cannot be beaten by specters.

Chapter 9

This Is Who I Am

The day I finished writing what I thought was the end of this book, Afghanistan fell to the Taliban. It felt necessary to say something.

Just because where we served fell to ruin does not mean what we did was worthless. I know a lot of us vets are struggling with feeling as though nothing we did in Afghanistan mattered. This sadly is true in terms of the lands and the way people there live. But the most important thing we need to remember is the value of life and the ones we saved in the moment. Similar to what I think firefighters feel when they can't get everyone out of a burning building, I feel great loss and failure that a nation is still under the rule of the undeserving. But we cannot control the will of others, let alone those a world away. We served and saved as best we could. This is what our service meant. Not that we saved a nation or changed the world, but that we did our best while we still had the power to stand in harm's way. The strong men and women of character we are doesn't falter because the house we grew up in was torn down. The stories we tell around a campfire that drop people's jaws still happened. The victories and terrible losses will always exist. Those who fell and the promises made in their memory will not be cast aside because of anger, fear, or hopelessness. I can't speak for all veterans, but I know my personally promised service contract expires only when I finally do.

Now settled into retirement, my productivity levels vary greatly day to day. There are days I do literally nothing but yard work, play my harmonica, and wrestle with my dog out back while we watch the sun crawl across the sky. Some days are full of continued physical and mental help from VA doctors and therapists. Many days have surprise visits from friends and family.

On rarer days, I'm blessed with purpose and still manage to get myself punched in the face. One New Year's Eve, I had been helping a friend tune up his motorcycle and was driving home to change clothes and shower before heading back for a small gathering to await the turn of the year. Halfway home, I came across a very fresh automobile accident. There was a white sedan facing the wrong direction in the middle of the intersection and smoke was emanating from its crumpled front end. Whatever had collided with the sedan was no longer present. My inability to ignore a situation forced me to pull into a bank parking lot on the corner and run to the car. To my astonishment, the car was still running with no driver or keys in the ignition. There was, however, a passed-out and bleeding man in the front passenger seat. The door was bent but not so badly it wouldn't open with a good-sized helping of applied effort. After the door begrudgingly opened, I used my pocketknife to cut the man out of his seatbelt and dragged him onto the street. He had a long, shallow cut on his head, but very little blood was coming out. He was breathing and moving, but not fully conscious. In the rush to get him out of what I thought was an imminent fireball, one of the man's shoes fell off and was still in the car. I held him in a sitting position when he realized a shoe had gone missing and assumed I had stolen it. The delirious man glanced a few times between his unshod foot and the stranger holding him.

Without much warning, he loudly proclaimed, "You stole my shoes!" and gave me the best right hook he could muster.

I fell backward and we both hit the pavement. A moment or two passed as I blankly gazed at the night sky. Once my brain rejoined my body, I sat up and rubbed my cheek. While pointing at his other foot,

I said, "You're only missing one. Your other shoe is in that car, about to catch fire. You want it, go get it."

The glaze in his eyes was still going strong as he tried to speak. "Oh God. I'm sor—I—the car—where is—" and he was out in la-la land again.

I propped him up so he wouldn't choke on vomit and went back to the car to see if I could turn it off. It died on its own with a loud clunk before I reached it. Someone had seen the wreck and phoned police. Emergency services arrived while I was walking back to check on Mr. Confused Fists of Fury. I explained to the officers what I'd been present for, and that if he asked about it, the man's shoe was on the floor of the passenger seat. A team of firemen doused the vehicle while officers directed traffic. After the situation was contained, the officers gave me permission to leave. I got in my rig and made my way out of the maze of emergency vehicles. As I passed the man, an ambulance medic was dressing his head. He was much more alert as he smiled sheepishly and mouthed, "Thank you. I'm sorry."

With windows down, I produced an ear-to-ear smile, waved kindly, and yelled at the top of my lungs, "Happy New Year! Don't worry about it. I kept your shoe!" and maniacally laughed my way home. Though I never did find out what happened with the driver, it was still a good day's work.

Every once in a while days turn into entire trips, and I get to volunteer with different veterans' groups. One such group is Flagpoles for a Cause. This particular charity obtains and installs flagpoles for Purple Heart recipients and Gold Star families. A Gold Star Family is a family whose soldier died in combat. Orton and Kendall are board members, so when a lot of work comes in, Orton drafts me for labor. Memorial Day weekend 2021 came around and the organization had what they call a "blitz." This is where four or five volunteers caravan out of state and install a handful of poles over one long weekend. For this blitz, we headed to Spokane, Washington. Though Spokane is where we stayed each night, we installed flagpoles in the surrounding counties with a few installs crossing the border into Idaho. The poles we were to install were almost entirely for guys I had served with in Ramadi fourteen years earlier. Some I hadn't seen since they were hurt.

Because we had so many flags to put up, there wasn't much time to talk or catch up at each brother's home. But just being able to physically check in and let them know I loved and missed them became one of the greatest rewards of the trip. As if it's an assigned extra duty, the well-being of those I served with is never far from my mind. Seeing how my brethren put their pieces back together also helps me stay focused on my way forward. The night before we returned home, Lissy came down and had dinner with us. That was the real treat. Being able to share a meal, joke, swap stories of progress, and observe moments of silence at the same table as Lissy and Orton was something uplifting that I can't fully express. It took nearly eight years for it to click, but I was finally able to see what Bill meant at the Cast and Blast. Though our service to the country formally ended, our service to each other has no expiration. If we choose it, our purpose is to make light the weight of others, sometimes simply by being in their presence. This was several days of excellent work.

That year also saw the ten-year anniversary of WALF forming, and I was asked to attend a Cast and Blast Alumni Reunion at Castle Valley Outdoors in the fall. This was a longer, five-day event focusing on the legacy portion of the name and how we've all grown, old guard and new. On day two, among others, I was asked to speak during dinner about my military experience and how WALF impacted my recovery and reintegration into civilian life. Without any time to prepare, the message was straight and true, but I feel my words may not have been clear enough. To know these fellow wounded soldiers and the other mentors involved with WALF has been beyond enlightening. The bonds I forged in war feel permanent and unbreakable but seeing those bonds in others thriving through five decades showed me proof of that permanence and gave me a deep well of hope.

Before I first attended back in 2012 I had already resolved to never take my own life, but the first time I was honestly able to imagine a future that didn't have to end early was witnessing the relationships of men who had gone through the same trials as me nearly fifty years earlier. I didn't get permission to use full names but along with Bill, I've also been mentored by Jim, Jon, and Terry. Each of these men are high-ranking, retired army officers and they are truly legendary. No permission for Bill Hansen's name either, but he's a WALF founder so he's stuck in the limelight anyhow. But my friends and family know their names and have heard me share some of their tales. If I ever find myself with children of my own, they will know who these men are, some of their stories, and how they inspire me to carry forward with honor.

I'll do just about anything for people I care about. Begrudgingly, I was even convinced to get ordained by one of those online ministries so I could officiate marriages. Young, a brother from Ramadi and Afghanistan, and his then-fiancée, Darcy, took me out to dinner one evening. They were celebrating their recent engagement and wanted to bring me along for some reason. Over appetizers, they informed me that I had a role to play. Young opened with a calm and casual demeanor.

"So, you already know we're getting married, and we're so glad..." he paused and glanced at Darcy so they could sync up to finish with a jubilant, "that you're officiating it!"

They may not have, but I clearly remember the inclusion of spirit fingers. Completely clueless on how to respond, "Uh, nope?" was all I could manage to say.

Before my brain had time to process any further objections, Darcy flopped a big, three-ring binder onto the table containing every wedding detail imaginable, including an entirely scripted sermon. I had been ambushed by well-prepared opponents. After several courses and hours, I relented and agreed to officiate their cottage dream wedding at Bear Lake. With every detail already set out for me, all I had to do was log in to the website, click a few things, and show up. Nothing could go wrong.

On the day of the wedding, everyone was going about their frantic pre-ceremony duties. It was peppy, mushy, nervous, chaotic hell. I did nothing but stay out of everyone's way, steal samples of wedding food I wasn't supposed to be eating yet, and hum nervously into my harmonica until the start of the ceremony. The cottage was inset from the lake and had an outside stone patio facing the water. We set up so bride and groom would face each other with Bear Lake forming a serene backdrop behind them. It was truly a picture-perfect place for a wedding. The thought of putting the sermon papers into a nice, leather-bound folder had crossed my mind. Instead, I opted to cut-and-paste the ceremony pages into an issue of the *Archie* comic book where he marries Veronica. Even took my time pasting the ceremony clippings over the actual word bubbles to preserve as much artwork as possible. Things were going great, and everyone got a kick out of me reading from a comic book. There was a multi-page spread of Archie and Veronica's ceremony in the middle of the comic that I particularly enjoyed. I skipped over it when pasting-in the ceremony thinking Darcy or Young might appreciate it too. I planned on giving it to them as a wedding present. Well, when I got to those pages, my TBI brain took a hard right turn and I forgot all about the real-life ceremony taking place. Alone in my own world, I began silently reading the comic book to myself, occasionally chuckling aloud. This went on for almost a minute before Darcy snapped me back to reality with a slap. As her hand collided with my neck and smashed my universe of one, she yell-whispered, "Ward! Hey!" Her hands and arms waved around furiously, showcasing where we were and the many people in attendance.

I gasped an audible, "Oh, no!" I fumbled forward to the nearest pasted paragraph and dropped right back into officiating like nothing happened. The three of us laughed our way through the vows and they had a great reception. I've officiated seven weddings to date, and they are always somehow full of hilarity and minor mishaps. I'm generally at fault, but people keep asking me to do it.

My oath to try and do better remains ever present. I still feel a tremendous amount of guilt about not answering Bull's final call for help. Nowadays, my entire family knows: Blood- or mud-born, we're stuck in this horseshit together. Faces wet with furiously salted tears or breath stolen by painful and forgotten memories, I'm always going to be around when someone needs to fall apart or go nuclear about how rough our existence sometimes is. In conversation, I'm often crude, loud, physically over animated, and pull very few verbal punches. When asked for my thoughts, if I don't understand all the facets of what's tearing someone apart, my words won't be of any help. If I see the turmoil was caused by their own selfish behaviors, my words will be unpleasantly honest. I think the reason veterans hide from the world when we come home is not because we no longer feel anything for others, but because we feel everything unrelentingly. Our experiences with war took away our desire to filter our mouths, sure, but they also took our ability to ignore what we perceive to be wrong. Toss in some brain damage and I no longer have the mental capacity to hold back or filter my thoughts. My blunt honesty has ended a handful of relationships but deepened my connection with others. I'm not perfect and my lack of filter or self-control has, on occasion, caused undue harm to people I care a lot about. I still have quite a few outstanding apologies to make. Emotional control is something I think we all have to strive to master. It means the difference between being able to tell someone they're fucking up while ensuring they know you still care and ending a decades-long relationship in bitter anguish. More works in progress.

A defense I sometimes encounter when trying to relate a moment from my service to someone's current hardship is: "You don't understand. I can't just do stuff like that. I'm not like you." Simply put, that is never true. Aside from being short, I'm a completely average human. I'm not highly educated, genetically enhanced, or mentally superior in any way. Hell, I'd place a wager that you are probably taller, smarter, and more graceful than I am. If my clumsy and disastrously inquisitive self can manage to bumble through harrowing life events, literally everyone can. Like pulling shrapnel out of your own forehead.

Back in 2007, after the big hit on Dusters, I was sitting on a bed in a hospital on Balad Air Base. Two doctors were standing a few feet away from me arguing about how deeply a piece of metal was lodged in my head. One fella thought it was subdermal (skin deep). The other guy thought it was sub-cranial (penetrated my skull). They weren't sure if they could take it out without causing further damage and were trying to decide the best machine to use to get an image. The shrapnel was metal so the option of an MRI, a giant electromagnet, was out. However, at that exact moment, none of what they were saying made any kind of sense to me. I stared and listened intently, but instead of actual words, the conversation sounded more like two cows chatting over a patch of grass. "Moo moo moooo? Moo mooooooo moo, moo mo mooo... Moo." They were yammering on, but it was doctor language that I simply could not grasp. The IED had left me mentally functioning at far below 100%. Doctors have their names for this condition; I call it jelly brain. Nothing makes sense and you aren't exactly a patient or understanding person while visiting this window-licker land. What I did know for certain was that my face hurt, and something was sticking out of it. I couldn't remember for certain, but I had a strong inkling that I, in fact, had not been born with a tiny, metallic unicorn horn on my face.

There happened to be one of those shiny, medical tool trays on a rolling table right next to the bed. You know, the table with the scalpel, scissors, and other fancy cutty- and grabby-type doctor tools. There was a pair of what I can only describe as the hospital version of needle-nose pliers laying amongst the medical tools. The doctors were way too involved arguing their different and massive medical knowledge and had no clue I was even there anymore. I reached my wobbly arm down and took the cold, metal tool into my hands so I could begin to decipher the secrets of how to use it. They were just pliers—fancy medical ones, sure—but simple pliers, nonetheless. Again, not a high-IQ day for me. After a moment of deliberation I decided I would operate this tiny device the same way you would with hedge trimmers, a handle in each hand. I gently flopped the tool onto my face and began to slide it, sort of side to side, slowly across my forehead. I was listening for a distinct *tink*-like sound. When I had used my jury-rigged metal-detection method enough to isolate the shrapnel's location, I closed the pliers around the foreign object as tightly as I could and pulled really hard, really fast. A piece of metal no bigger than my pinkie nail came right out. As my face began running red with blood, I victoriously looked up to the two arguing bovine and presented pliers in one hand and my itty bitty unicorn horn in the other. With a doe-eyed, clueless smile, I cheerfully proclaimed, "It's gone! This thing worked great." All that time mooing over nothing and I helped in seconds. The doctors were not pleased. With looks of mild panic and disbelief from multiple scrub-adorned humans, I was rushed in to get a CT scan of my head.

In reality, I was a jelly-brained and impatient idiot who simply tried an idea that happened to work out. But that's my point: Deeds people often see as heroic or unbelievable aren't generally viewed that way by the people who do them. The reasons are never the same from one veteran to the next, but just about all of us would never use the word hero when describing ourselves or what we've done in war. There's always a few, but those folks are generally more full of shit than a septic tank. If you had our jobs, our training, our same people at your side—you know, really stepped into our lives—you probably would have done your own very similar version of whatever it was that we did. In my experience, no one is any good at accurately imagining living someone else's life simply because we've only ever existed as ourselves.

Every moment in this book, and most of my life really, has been just like the shrapnel. When they occurred, I simply did whatever I could to make the outcome of each situation the best possible one. Sometimes I was right on the money. Other occasions I dropped the ball completely and, at best, whole missions were scrubbed. But I never stopped trying, nor will I. That's all any of this book is about. To remind anyone who reads it, especially myself, that each of us is more than what we currently see in a mirror. No matter what stage of life we are in, what we see today only represents us right now, today. Sure, some of our trials and hardships are visible. But nowhere near all of what we are, with our countless lessons learned and overly developed peculiarities, can be seen in a simple reflection. The unseen is a vast ocean of knowledge that can help others if we learn to share the ugly moments we've survived. When I try to explain this concept is when I sometimes get the, "Well, I can't imagine ever doing what you did," defense. *Honestly?* Me either. Not in a million years would I have ever imagined I would someday remove metal, by hand, from my own forehead. But in the moment it made perfect sense. If it had been you, you might have come to the same conclusion as me. Who knows, maybe you would have done it without using pliers at all. It only seems crazy because you didn't see it firsthand. If you had, you would laugh at me every time you thought of unicorns. Hopefully you will now.

How and if someone recovers from awful life events will be different for each person, but it always boils down to choice. At one point or another we all mentally, emotionally, and physically get our teeth kicked in by life itself. Most folks more than once. Even though "it" could be anything and you won't see it coming, your life will be effortlessly demolished. When that happens, you can wallow in defeat, riddled with guilt, whipping your own back until you yourself have become useless. Or you can do the simplest thing and stand back up. No plan to charge forward, no plan to crawl backward amid all the destruction, just stand. It will never be easy. After being beaten down, standing is probably one of the hardest paths to take, but it is the simplest. Show everyone, yourself included, that no matter how beaten you are, you remain unbroken and stand firm in your defiance of failure. Slowly, as strength returns, begin moving forward and rebuilding. Believe me, nothing gets fixed laying in the dirt with blood and tears on your face.

In the end, I simply want to live in a manner worthy of those who've sacrificed for me. There is still an immeasurable amount of work in the areas of self-improvement and universe beautification—more than I can accomplish in my remaining lifetime, in fact. Thankfully, I've been given enough stubborn strength to try anyway.

The End

About the Author

Ward Mayfield is a retired Army Staff Sergeant who is weirdly uncomfortable with the realization that authors write their own bios in 3rd person. He served in the 321st Engineer Battalion as a Combat Engineer, completing over 300 missions during three tours. Those tours took him in Iraq twice and Afghanistan once. His primary job was route clearance — the detection and removal of IEDs from public roadways. Among the awards he has received are two Purple Heart Medals, an Army Commendation for Valor, and a Bronze Star. Ward is currently living out his retirement in the greater Salt Lake area.

About the Publisher

Tactical 16 Publishing is an unconventional publisher that understands the therapeutic value inherent in writing. We help veterans, first responders, and their families and friends to tell their stories using their words.

We are on a mission to capture the history of America's heroes: stories about sacrifices during chaos, humor amid tragedy, and victories learned from experiences not readily recreated — real stories from real people.

Tactical 16 has published books in leadership, business, fiction, and children's genres. We produce all types of works, from self-help to memoirs that preserve unique stories not yet told.

You don't have to be a polished author to join our ranks. If you can write with passion and be unapologetic, we want to talk. Go to Tactical16.com to contact us and to learn more.

TACTICAL 16
PUBLISHING

Over the years, friends and family have asked what living through combat in Iraq and Afghanistan and the resulting injuries is really like. My answer is usually something dismissive, like "it was mostly a pleasure," or "absolute lunacy," and a casual smile.

While true, I've responded that way because to detail war so that the person I am speaking with truly understands what it was like for me is incredibly painful. But I wanted to share myself with those closest to me and needed to try to get my most personally influential experiences and resulting perspective changes down on paper.

This is my take on some of the lessons good 'ole Uncle Sam gave me, and how I continually choose to live with them. So even if my brain injuries take over at some point down the road, at least future me will have a reference to know what the hell happened.

Perhaps it will help friends and family or maybe even help total strangers understand just a small example of what survivors of war look like on the inside.

TACTICAL16
PUBLISHING
www.TACTICAL16.com

Made in the USA
Monee, IL
20 October 2025

32438336R00074